Vertebrates

Animal Types 2

Vertebrates

Maureen A. Robinson and Judith F. Wiggins

HUTCHINSON EDUCATIONAL

HUTCHINSON EDUCATIONAL LTD
3 Fitzroy Square London W1

London Melbourne Sydney
Auckland Johannesburg Cape Town
and agencies throughout the world

First published 1971

B/591

*This book has been cold typeset in Univers by Design Practitioners, Sevenoaks
printed in Great Britain by Anchor Press, and bound by Wm. Brendon,
both of Tiptree, Essex.*

ISBN 0 09 108780 5 (cased)
     0 09 108781 3 (paper)

iv

Contents

Contents

Preface

We hope this book serves the same purpose as Animal Types I in that it provides a clear, simple classification, drawings and photographs of external features and further annotated drawings of points of interest from the wide range of animal types in the Protochordates and Vertebrates. As with the first volume the drawings are of actual specimens and are not intended for direct reproduction by the students. The drawings are not to scale as they are intended for use in the laboratory alongside specimens, and to supplement visits to museums and zoos.

The section "Experimental Approach to the Vertebrates" was contributed by Miss Rosemary Richardson, lecturer in Biology and Zoology at Walbrook College. This section deals with simple significant experiments which may be carried out by students in the laboratory to supplement their study of Vertebrate Types.

The authors would like to thank the Zoology Department of the Natural History Section of the British Museum for reading and criticising the manuscript, for access to specimen collections and literature, and also the Science Department of Walbrook College for the use of specimens.

Acknowledgement is due to the following for permission to use illustrations: Newman's Natural History Photographic Agency, Page 46; Trustees of the British Museum (Natural History), Pages 33, 48, 51, 53, 57, 60, 62, 71; Zoological Society of London, Pages 30, 39, 40, 54, 65, 68.

Kingdom Animalia

SUBKINGDOM METAZOA

Animals consisting of tissues and organs.

GRADE BILATERIA

Animals exhibiting bilateral symmetry, with tissues, definite organs and true mesoderm.

Subgrade: Coelomata

Animals with a space—true body cavity or coelom, lined by peritoneum of mesodermal origin.

Superphylum Deuterostomia

Enterocoelomates—animals in which a true coelom arises from pouches enclosing portions of archenteron.

PHYLUM CHORDATA

Ubiquitous distribution.
Usually solitary and motile.
Triploblastic, metamerically segmented coelomates.
Bilaterally symmetrical.
Development direct or via a tailed larva.

External Features

Pharynx pierced by a number of visceral clefts or gill slits.
Post-anal tail present at some stage of development.

Internal Features

Notochord—axial skeleton present ventral to nerve cord, may be replaced by bone or cartilage to form a vertebral column in higher chordates.
Central nervous system—dorsal tubular nerve cord, gives off paired segmental nerves present at some stage of development.
Brain—three lobed, modified anterior end of nerve cord present in higher chordates.
Closed blood system—blood flows backwards in dorsal vessel, forwards in ventral vessel.
Endostyle present, forms thyroid gland in higher chordates.

GENERAL REFERENCES

Chapman, G. and Barker, W.B.	1964	Zoology	London
Larousse	1967	Encyclopaedia of Animal Life	London
Romer, A.S.	1968	Man and the Vertebrates	London
Saunders, J.T. and Manton, S.M.	(Fourth edition revised by Manton S.M. and Brown M.E.) 1969 Manual of Practical vertebrate Morphology		Oxford
Savory, T.H.	1968	Introduction to Zoology	London
Vallim, J.	1966	Animal Kingdom	New York
Vines, A.E. and Rees, N.	1959	Plant and Animal Biology	London
Weisz, P.B.	1961	Elements of Biology	New York
Yapp, W.B.	1912	Borradailes Manual of Elementary Zoology	London
Young, J.Z.	1950	Life of Vertebrates	London

I Subphylum Urochordata (Tunicata)

Aquatic—marine.
Sessile or free swimming.
Often colonial through budding.

External Features

Body sac like.
Tunic or test of tunicin (related to cellulose).
Head end absent.
Metameric segmentation not obvious.
Post anal tail present in larva only (anus opens to the atrium).

Internal Features

Dorsal atrium present in adult.
Dorsal nerve cord present in larva—reduced to ganglion in adult.
Notochord present in larva—absent in adult.
Numerous gill clefts for filter feeding and respiration.
Blood flow backwards and forwards in ventral vessel—numerous blood spaces present.
Tubular excretory organs absent.
Endostyle present.

CLASS ASCIDIACEA

Sessile, solitary or colonial.
Ascidian tadpole larva exhibits chordate characters—lost at metamorphosis.
e.g. *Ciona*—sea squirt

REFERENCE FOR CLASS ASCIDIACEA

Barrington, E.J.W.   1965   Biology of the Hemichordata
and Protochordata   London

Ciona
Sea squirt

PHYLUM CHORDATA

SUBPHYLUM UROCHORDATA [TUNICATA]

CLASS ASCIDIACEA

GENUS *Ciona*

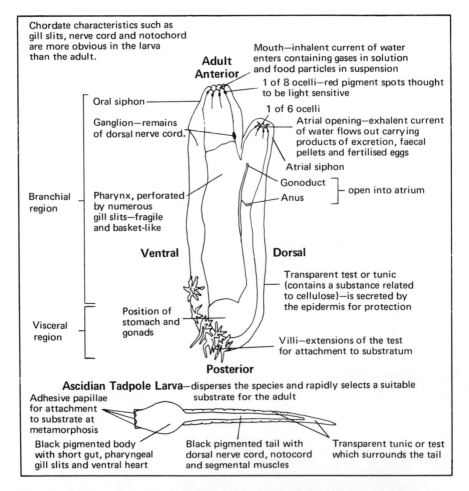

Chordate characteristics such as gill slits, nerve cord and notochord are more obvious in the larva than the adult.

Adult Anterior

Mouth—inhalent current of water enters containing gases in solution and food particles in suspension

1 of 8 ocelli—red pigment spots thought to be light sensitive

Oral siphon

1 of 6 ocelli

Ganglion—remains of dorsal nerve cord.

Atrial opening—exhalent current of water flows out carrying products of excretion, faecal pellets and fertilised eggs

Atrial siphon

Gonoduct — open into atrium

Branchial region

Pharynx, perforated by numerous gill slits—fragile and basket-like

Anus

Ventral

Dorsal

Transparent test or tunic (contains a substance related to cellulose)—is secreted by the epidermis for protection

Visceral region

Position of stomach and gonads

Villi—extensions of the test for attachment to substratum

Posterior

Ascidian Tadpole Larva—disperses the species and rapidly selects a suitable substrate for the adult

Adhesive papillae for attachment to substrate at metamorphosis

Black pigmented body with short gut, pharyngeal gill slits and ventral heart

Black pigmented tail with dorsal nerve cord, notocord and segmental muscles

Transparent tunic or test which surrounds the tail

The sea squirt is usually solitary and sessile. It is found in shallow sea water, attached to rocks, stones and undersides of piers, buoys and ships. It feeds on diatoms and organic particles sifted from the sea water by its ciliary filter feeding mechanism.

The sea squirt is hermaphrodite—fertilised eggs are liberated during the annual breeding season and hatch into ascidian tadpole larvae which are free swimming, but do not feed. Five hours after hatching, each larva settles, the tail and notochord are lost, the nerve cord is reduced and metamorphosis takes place resulting in the adult form, which periodically can be seen to squirt out jets of water from both the atrial and oral siphons. This is due to rapid contraction of the muscle fibres of the body wall which force sea water back through the gill slits to free them of any trapped food particles and sand.

4

II Subphylum
Cephalochordata (Acrania)

Aquatic—marine.

Free swimming and burrowing.

External Features

Body fish like.

Head end present but little specialisation.

Post-anal tail.

Internal Features

Metameric segmentation of myotomes and gonads obvious.

Atrium present.

Dorsal nerve cord present.

Notochord along whole length of body persists throughout life.

Numerous gill clefts for filter feeding and respiration.

Blood flow backwards in dorsal vessel, forwards in ventral vessel.

Excretion by ectodermal protonephridia.

Endostyle present.

e.g. *Branchiostoma*—lancelet or amphioxus

REFERENCE FOR SUBPHYLUM CEPHALOCHORDATA

Barrington, E.J.W.   1965   Biology of the Hemichordata
                                 and Protochordata   London

Branchiostoma
Lancelet or Amphioxus

PHYLUM CHORDATA

SUBPHYLUM CEPHALOCHORDATA

GENUS *Branchiostoma*

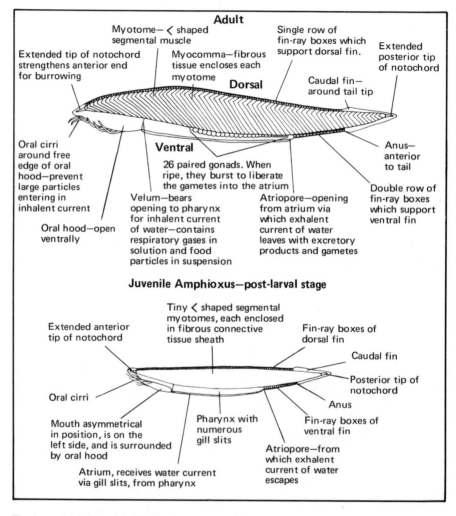

Adult

Myotome—< shaped segmental muscle

Extended tip of notochord strengthens anterior end for burrowing

Myocomma—fibrous tissue encloses each myotome

Dorsal

Single row of fin-ray boxes which support dorsal fin.

Extended posterior tip of notochord

Caudal fin— around tail tip

Oral cirri around free edge of oral hood—prevent large particles entering in inhalent current

Ventral

26 paired gonads. When ripe, they burst to liberate the gametes into the atrium

Anus— anterior to tail

Double row of fin-ray boxes which support ventral fin

Oral hood—open ventrally

Velum—bears opening to pharynx for inhalent current of water—contains respiratory gases in solution and food particles in suspension

Atriopore—opening from atrium via which exhalent current of water leaves with excretory products and gametes

Juvenile Amphioxus—post-larval stage

Extended anterior tip of notochord

Tiny < shaped segmental myotomes, each enclosed in fibrous connective tissue sheath

Fin-ray boxes of dorsal fin

Caudal fin

Posterior tip of notochord

Oral cirri

Anus

Mouth asymmetrical in position, is on the left side, and is surrounded by oral hood

Pharynx with numerous gill slits

Fin-ray boxes of ventral fin

Atriopore—from which exhalent current of water escapes

Atrium, receives water current via gill slits, from pharynx

The lancelet is found in shallow sea water. It burrows in the sand by day and swims in the surface water by night to feed on organic particles and small plankton, which it sifts from the sea water by its ciliary feeding mechanism.

The sexes are separate. Breeding occurs in early summer when gametes are shed into the surface waters. Fertilised eggs develop into extreme asymmetrical larvae which feed and swim in the surface waters for six months. Then they metamorphose into the adult form and leave the surface waters to burrow in the sand.

The lancelet is caught in great numbers off the coast of Hong Kong where it is considered to be a delicacy.

III Subphylum
Vertebrata (Craniata)

External Features

Head obvious and well developed.

Visceral clefts are restricted in number and often lost in the adult. Primarily concerned with respiration.

Two pairs of limbs usually present.

Internal Features

Notochord of embryonic stage wholly or partially replaced by a vertebral column.
Nervous system well developed with a definite brain encased by a cranium.
Heart—well developed and muscular.
Blood vascular system—closed.
Kidneys—organs of nitrogenous excretion.
Sexes usually separate.

Class Chondrichthyes (Selachii)

Cartilaginous fish.
Poikilothermic.
Mainly marine.

External Features
Body fusiform or depressed.
Exoskeleton of placoid scales or dermal denticles.
Spiracle and gill clefts not covered by an operculum or gill cover in sharks and rays.
External nares and mouths on the ventral surface of the head.
Pectoral and pelvic fins, paired limbs—large.
Median fins present.
Heterocercal tail fin—tail muscular and forms main propulsive organ except in rays.
Cloaca present.

Internal Features
Endoskeleton—cartilaginous.
Gills—respiratory organs.
Spiral valve present in intestine.
Heart—one auricle, one ventricle. Pumps only deoxygenated blood.
Inner ear present.
Swim bladder or lung absent.

Fertilisation is internal. Claspers, the male intromittent organs, transfer sperm to the female. Large yolky egg enclosed in a horny capsule laid in small numbers. Numerous species of sharks and rays are ovoviviparous or viviparous.
e.g. *Scyliorhinus*—dogfish
   Raja—skate or ray

REFERENCES FOR CLASS CHONDRICHTHYES

Kennedy, M.	1969	The Sea Angler's Fishes	London
Lanham, U.	1962	The Fishes	New York
Marshall, N.B.	1965	The Life of Fishes	London
Wheeler, A.	1969	The Fishes of the British Isles and North West Europe	London

Suggested films for Class Chondrichthyes

| G.B. Instructional | Dogfish as a Vertebrate | B/W 10 mins. |
| Rank Film Library | Fish | Colour 11 mins. |

Scyliorhinus
Dogfish

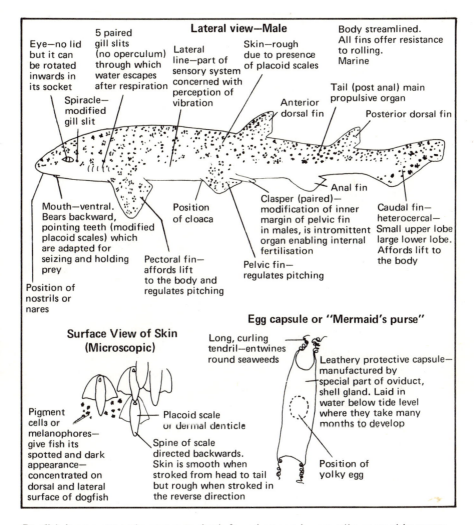

Lateral view—Male

Eye—no lid but it can be rotated inwards in its socket

5 paired gill slits (no operculum) through which water escapes after respiration

Lateral line—part of sensory system concerned with perception of vibration

Skin—rough due to presence of placoid scales

Body streamlined. All fins offer resistance to rolling. Marine

Spiracle—modified gill slit

Anterior dorsal fin

Tail (post anal) main propulsive organ

Posterior dorsal fin

Mouth—ventral. Bears backward, pointing teeth (modified placoid scales) which are adapted for seizing and holding prey

Position of cloaca

Anal fin

Clasper (paired)—modification of inner margin of pelvic fin in males, is intromittent organ enabling internal fertilisation

Caudal fin—heterocercal—Small upper lobe large lower lobe. Affords lift to the body

Pectoral fin—affords lift to the body and regulates pitching

Pelvic fin—regulates pitching

Position of nostrils or nares

Surface View of Skin (Microscopic)

Pigment cells or melanophores—give fish its spotted and dark appearance—concentrated on dorsal and lateral surface of dogfish

Placoid scale or dermal denticle

Spine of scale directed backwards. Skin is smooth when stroked from head to tail but rough when stroked in the reverse direction

Egg capsule or "Mermaid's purse"

Long, curling tendril—entwines round seaweeds

Leathery protective capsule—manufactured by special part of oviduct, shell gland. Laid in water below tide level where they take many months to develop

Position of yolky egg

Dogfish is an extremely common shark found on sandy, gravelly or muddy coasts where it dwells on the sea bottom. It feeds on molluscs, crustaceans and small fish, which are swallowed whole.

Mating is carried out in shallow water, and atypically of fish, fertilisation is internal. Sperm are transferred from the male to the female by means of a pair of claspers. Egg laying occurs all the year round in shallow to quite deep water. The eggs are protected by an egg case from which the young hatch when they are about four inches in length.

Dogfish is of economic importance as food—rock salmon—and as an animal for dissection purposes.

Scyliorhinus
Dogfish

PHYLUM CHORDATA

SUBPHYLUM VERTEBRATA [CRANIATA]

CLASS CHONDRICHTHYES [SELACHII]

GENUS *Scyliorhinus*

Embryo (removed from egg case)

Stalk of
yolk sac

Eye

External gills—long
vascular threads
project from gill
clefts

Yolk sac attached
to belly of young
fish

Developing fins

Dorsal fin fold

Young dogfish hatch at six to twelve months. When young, it is protected by the egg case and fed on yolk. It is well developed at birth and resembles the adult.

Vertical section through skin

Placoid scale

Enamel—hard, transparent protective
coat, secreted by epidermis

Posterior

Dentine—contains fine canals which
carry blood and nerve supply. Secreted
by odontoblasts in pulp cavity

Backward
pointing spine

Pulp cavity—contains connective
tissue, nerve and blood supply

Epidermis—stratified
epithelial tissue

Dermis—connective
tissue with fibres
arranged at right angles.
Gives great strength
and flexibility

Mucous gland

Basal plate

Pigment cells or
chromatophores—
afford animal
protective colouring

Developing placoid scale—
continual replacement of scales

Underlying
muscle (striped)

PHYLUM CHORDATA

SUBPHYLUM VERTEBRATA [CRANIATA]

CLASS CHONDRICHTHYES [SELACHII]

GENUS *Raja*

Raja
Skate or Ray

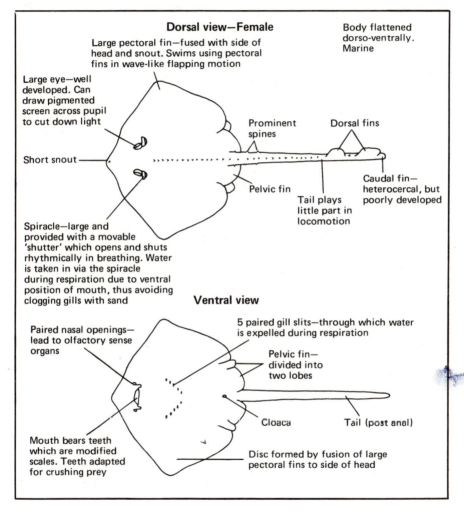

Dorsal view—Female

Large pectoral fin—fused with side of head and snout. Swims using pectoral fins in wave-like flapping motion

Body flattened dorso-ventrally. Marine

Large eye—well developed. Can draw pigmented screen across pupil to cut down light

Prominent spines

Dorsal fins

Short snout

Pelvic fin

Caudal fin— heterocercal, but poorly developed

Tail plays little part in locomotion

Spiracle—large and provided with a movable 'shutter' which opens and shuts rhythmically in breathing. Water is taken in via the spiracle during respiration due to ventral position of mouth, thus avoiding clogging gills with sand

Ventral view

Paired nasal openings— lead to olfactory sense organs

5 paired gill slits—through which water is expelled during respiration

Pelvic fin— divided into two lobes

Cloaca

Tail (post anal)

Mouth bears teeth which are modified scales. Teeth adapted for crushing prey

Disc formed by fusion of large pectoral fins to side of head

The terms "skate" and "ray" are more or less synonymous. Skate is applied to the large, long-nosed members and ray to the smaller, snub-nosed ones.

They are bottom dwellers found on mud, sand and gravel in shallow to deep water. Their food consists of crustaceans, small fish and worms which are located by smell. The fish are caught by swimming upward and enveloping them with the pectoral fins.

The skate or ray migrate inshore for mating during which sperm are transferred from the male to the female by means of the claspers (modified pelvic fins in male). Large, yolky eggs are laid within egg capsules or Mermaid's purses close to the shore. The young fish are about 3 inches long on hatching.

11

Class Osteichthyes (Pisces)

Bony fish.
Poikilothermic.
Fresh water and marine.

External Features

Body usually fusiform.
Exoskeleton of flat bony scales generally present.
Operculum or gill cover protecting gills.
External nares usually dorsal, mouth usually terminal.
Pectoral and pelvic fins, paired limbs—small.
Median fins present.
Homocercal tail fin usually present—tail muscular and forms main propulsive organ.
Cloaca not usually present. Anus and urinogenital apertures distinct.

Internal Features

Endoskeleton—bony.
Gills—respiratory organs.
Spiral valve—very seldom present in intestine.
Heart—one auricle, one ventricle. Pumps only deoxygenated blood.
Inner ear present.
Swim bladder or lung present—usually modified as hydrostatic organ.

Fertilisation usually external.
In most species numerous, small eggs are laid with little yolk.
Young often different from adults—known as larvae.
e.g. *Clupea*—herring

REFERENCES FOR CLASS OSTEICHTHYES

Kennedy, M.	1969	The Sea Angler's Fishes	London
Lanham, U.	1962	The Fishes	New York
Marshall, N.B.	1965	The Life of Fishes	London
Wheeler, A.	1969	The Fishes of the British Isles and North West Europe	London

Suggested film for Class Osteichthyes

Rank Film Library	Fish	Colour	11 mins.

PHYLUM CHORDATA

SUBCLASS VERTEBRATA [CRANIATA]

CLASS OSTEICHTHYES [PISCES]

GENUS *Clupea*

Clupea
Herring

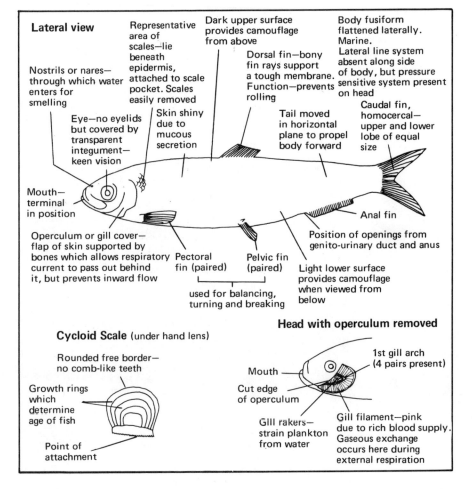

Lateral view

Representative area of scales—lie beneath epidermis, attached to scale pocket. Scales easily removed

Dark upper surface provides camouflage from above

Body fusiform flattened laterally. Marine.

Lateral line system absent along side of body, but pressure sensitive system present on head

Nostrils or nares— through which water enters for smelling

Dorsal fin—bony fin rays support a tough membrane. Function—prevents rolling

Eye—no eyelids but covered by transparent integument— keen vision

Skin shiny due to mucous secretion

Tail moved in horizontal plane to propel body forward

Caudal fin, homocercal— upper and lower lobe of equal size

Mouth— terminal in position

Operculum or gill cover— flap of skin supported by bones which allows respiratory current to pass out behind it, but prevents inward flow

Pectoral fin (paired)

Pelvic fin (paired)

used for balancing, turning and breaking

Anal fin

Position of openings from genito-urinary duct and anus

Light lower surface provides camouflage when viewed from below

Cycloid Scale (under hand lens)

Rounded free border— no comb-like teeth

Growth rings which determine age of fish

Point of attachment

Head with operculum removed

1st gill arch (4 pairs present)

Mouth

Cut edge of operculum

Gill rakers— strain plankton from water

Gill filament—pink due to rich blood supply. Gaseous exchange occurs here during external respiration

The herring is a pelagic fish found off the coast of Great Britain. There are several races of herring which differ from one another in both morphological and biological ways.

Herrings are plankton feeders in their young stages. The plankton is strained from the current of water passing over the gills by means of the gill rakers. In later life it will eat crustaceans and small fish.

Herrings of different races spawn at different times, each race having distinct spawning grounds. Fertilisation is external with the females laying eggs in vast quantities. The eggs, which are heavier than sea water, sink and adhere to the sea bed. The young larvae hatch out after two weeks and remain in shallow water near to the spawning ground. After two years they gradually move to deeper water.

Herrings are of economic importance as food.

Clupea
Herring

PHYLUM CHORDATA

SUBPHYLUM VERTEBRATA [CRANIATA]

CLASS OSTEICHTHYES [PISCES]

GENUS *Clupea*

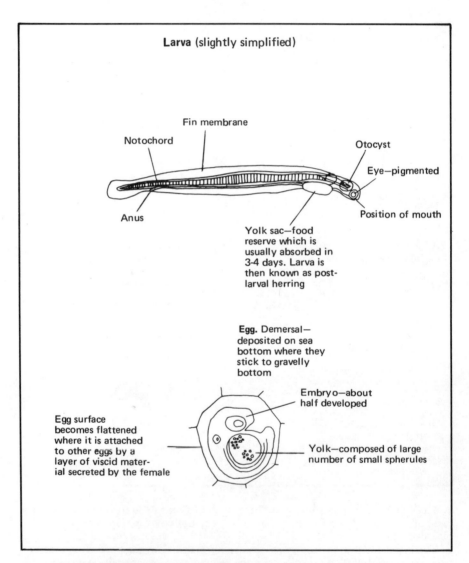

Larva (slightly simplified)

Fin membrane

Notochord

Otocyst

Eye—pigmented

Anus

Position of mouth

Yolk sac—food reserve which is usually absorbed in 3-4 days. Larva is then known as post-larval herring

Egg. Demersal—deposited on sea bottom where they stick to gravelly bottom

Embryo—about half developed

Egg surface becomes flattened where it is attached to other eggs by a layer of viscid material secreted by the female

Yolk—composed of large number of small spherules

The egg hatches in 1-2 weeks, and the larval stage lasts for 3-4 days. These two stages in the life of the herring occur near the sea bottom. The post-larval herring leaves the bottom and migrates into mid-water in its search for food and then on into surface and following this, littoral waters. The herring attains sexual maturity at a length of 8-9 inches and this is probably during the third year.

Larva after Entenbaum's *Nordische Plankton*
Egg after Arthur Thomas Masterman 1897, *Life Histories of British Marine Food Fishes.*

Class Amphibia

Poikilothermic

Terrestrial, arboreal, aquatic (fresh water to brackish but never marine).

External Features

Scales usually absent from the skin.

Modified pentadactyl limbs—only four or fewer digits on the forelimbs.

Tympanum or ear drum often visible on surface of skin.

Internal Features

Ear consists of an inner and middle chamber.

Lungs usually present in the adult.

Heart consists of two auricles, one ventricle for passage of oxygenated and deoxygenated blood.

Adult excretory organs are usually mesonephric kidneys—excrete urea.

Cloaca present receives ureters, genital ducts and rectum.

Fertilisation usually external.

Usually lay eggs in water.

Larvae almost invariably aquatic until metamorphosis.

External gills present in some larvae.

Orders

ANURA (Salientia)

1. Tail lost at metamorphosis
2. Vertebral column short.
3. Forelimbs short, hindlimbs long.
4. Gill slits close in adult.

    e.g. *Rana*—frog
        Bufo—toad

URODELA (Caudata)

1. Tail persists throughout life.
2. Vertebral column long.
3. Fore- and hindlimbs short, almost equal in length—sometimes completely absent.
4. External and internal gills may persist in a few adults.

    e.g. *Triturus*—newt
        Salamandra—salamander
        Ambystoma—axolotl

REFERENCES FOR CLASS AMPHIBIA

Cochran, D.M.	1961	Living Amphibia of the World	London
Noble, G.K.	1931	Biology of Amphibia	New York
Savage, M.	1961	Ecology and Life History of Common Frog	London
Smith, M.A.	1951	The British Amphibians and Reptiles	London
Steward, J.	1969	Tailed Amphibia of Europe	London

Suggested Films for Class Amphibia

Films Assoc. of California	Amphibians (Frogs, Toads and Salamanders)	B/W	11 mins.
G.B. Instructional	Life of the Newt	B/W	10 mins.
G.B. Instructional	Life History of the Frog	B/W	17 mins.
G.B. Instructional	Introduction to the Frog	B/W	18 mins.

Rana
Frog

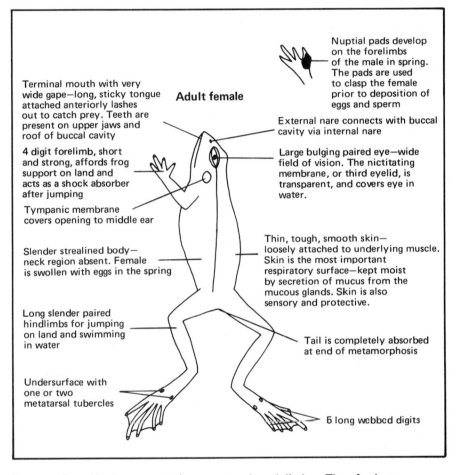

Nuptial pads develop on the forelimbs of the male in spring. The pads are used to clasp the female prior to deposition of eggs and sperm

Terminal mouth with very wide gape—long, sticky tongue attached anteriorly lashes out to catch prey. Teeth are present on upper jaws and roof of buccal cavity

Adult female

External nare connects with buccal cavity via internal nare

4 digit forelimb, short and strong, affords frog support on land and acts as a shock absorber after jumping

Large bulging paired eye—wide field of vision. The nictitating membrane, or third eyelid, is transparent, and covers eye in water.

Tympanic membrane covers opening to middle ear

Slender strealined body— neck region absent. Female is swollen with eggs in the spring

Thin, tough, smooth skin— loosely attached to underlying muscle. Skin is the most important respiratory surface—kept moist by secretion of mucus from the mucous glands. Skin is also sensory and protective.

Long slender paired hindlimbs for jumping on land and swimming in water

Tail is completely absorbed at end of metamorphosis

Undersurface with one or two metatarsal tubercles

5 long webbed digits

Frogs are found in damp vegetation near ponds and ditches. They feed on various invertebrates including worms, slugs and flies. In temperate regions frogs hibernate, usually in the mud of ponds or ditches. When they emerge the male croaks to attract the female, who is swollen with eggs, to the pond. The male mounts the back of the female and they swim around in this position —amplexus— for several days until the eggs and sperm are deposited. Emergence from the egg and metamorphosis are effected by temperature. When the tadpole larva emerges, it hangs onto weeds by its cement gland, feeds on yolk in its yolk sac and respires via external gills. A mouth with horny jaws and teeth develops enabling the tadpole to feed on water plants. The external gills shrivel and internal gills now function. Later fore- and hindlimbs are fully formed, feeding ceases and the skin is shed. After metamorphosis a young frog with a stumpy tail leaves the pond. Frogs may take up to four years to reach maturity.

Rana
Frog

PHYLUM CHORDATA

SUBPHYLUM VERTEBRATA [CRANIATA]

CLASS AMPHIBIA

ORDER ANURA [SALIENTIA]

GENUS *Rana*

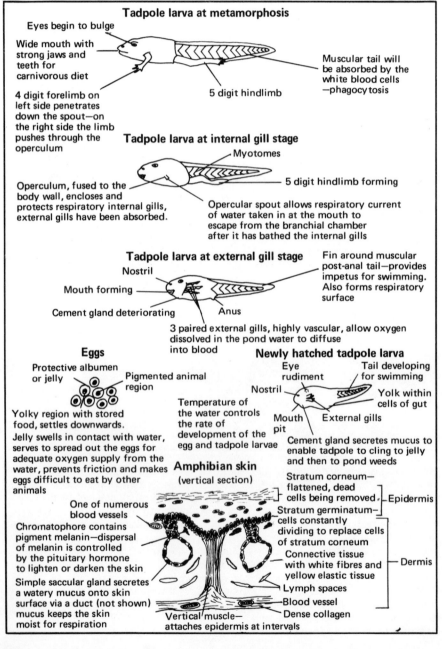

Tadpole larva at metamorphosis

Eyes begin to bulge

Wide mouth with strong jaws and teeth for carnivorous diet

Muscular tail will be absorbed by the white blood cells —phagocytosis

4 digit forelimb on left side penetrates down the spout—on the right side the limb pushes through the operculum

5 digit hindlimb

Tadpole larva at internal gill stage

Myotomes

5 digit hindlimb forming

Operculum, fused to the body wall, encloses and protects respiratory internal gills, external gills have been absorbed.

Opercular spout allows respiratory current of water taken in at the mouth to escape from the branchial chamber after it has bathed the internal gills

Tadpole larva at external gill stage

Nostril

Fin around muscular post-anal tail—provides impetus for swimming. Also forms respiratory surface

Mouth forming

Cement gland deteriorating

Anus

3 paired external gills, highly vascular, allow oxygen dissolved in the pond water to diffuse into blood

Eggs

Protective albumen or jelly

Pigmented animal region

Yolky region with stored food, settles downwards.

Jelly swells in contact with water, serves to spread out the eggs for adequate oxygen supply from the water, prevents friction and makes eggs difficult to eat by other animals

Temperature of the water controls the rate of development of the egg and tadpole larvae

Newly hatched tadpole larva

Eye rudiment

Tail developing for swimming

Nostril

Yolk within cells of gut

Mouth pit

External gills

Cement gland secretes mucus to enable tadpole to cling to jelly and then to pond weeds

Amphibian skin

(vertical section)

One of numerous blood vessels

Chromatophore contains pigment melanin—dispersal of melanin is controlled by the pituitary hormone to lighten or darken the skin

Simple saccular gland secretes a watery mucus onto skin surface via a duct (not shown) mucus keeps the skin moist for respiration

Stratum corneum— flattened, dead cells being removed

Epidermis

Stratum germinatum— cells constantly dividing to replace cells of stratum corneum

Connective tissue with white fibres and yellow elastic tissue

Dermis

Lymph spaces

Blood vessel

Dense collagen

Vertical muscle— attaches epidermis at intervals

PHYLUM CHORDATA

SUBPHYLUM VERTEBRATA [CRANIATA]

CLASS AMPHIBIA

ORDER ANURA [SALIENTIA]

GENUS *Bufo*

Bufo
Toad

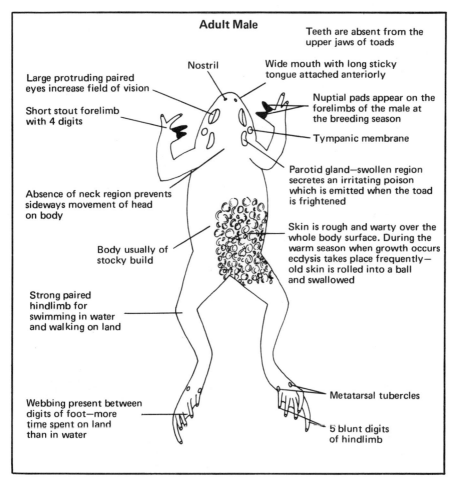

Adult Male

Teeth are absent from the upper jaws of toads

Nostril

Wide mouth with long sticky tongue attached anteriorly

Large protruding paired eyes increase field of vision

Nuptial pads appear on the forelimbs of the male at the breeding season

Short stout forelimb with 4 digits

Tympanic membrane

Parotid gland—swollen region secretes an irritating poison which is emitted when the toad is frightened

Absence of neck region prevents sideways movement of head on body

Skin is rough and warty over the whole body surface. During the warm season when growth occurs ecdysis takes place frequently— old skin is rolled into a ball and swallowed

Body usually of stocky build

Strong paired hindlimb for swimming in water and walking on land

Metatarsal tubercles

Webbing present between digits of foot—more time spent on land than in water

5 blunt digits of hindlimb

Toads live in a wide variety of relatively dry habitats. The British toad lives in woods and fields, and in very dry conditions it aestivates in moisture-preserving cracks. In winter it hibernates. Toads use their mobile sticky tongues to catch their prey—invertebrates such as insects—which are often injurious to crops.

Breeding occurs in the spring, when the male sings for several days to attract the female to a shallow pond or ditch. The male mounts the back of the female and clasps her with his forelimbs. They swim in this position for several days until the female deposits thousands of eggs laid in strings, the male depositing sperm simultaneously. Tadpole larvae with external gills emerge from the jelly and feed on algae. Later fore- and hindlimbs appear, the lungs function, the tail is slowly reabsorbed, and they feed on invertebrates. When metamorphosis is complete, the young toads leave the pond but they will return to it to breed when mature.

Bufo
Toad

PHYLUM CHORDATA

SUBPHYLUM VERTEBRATA [CRANIATA]

CLASS AMPHIBIA

ORDER ANURA [SALIENTIA]

GENUS *Bufo*

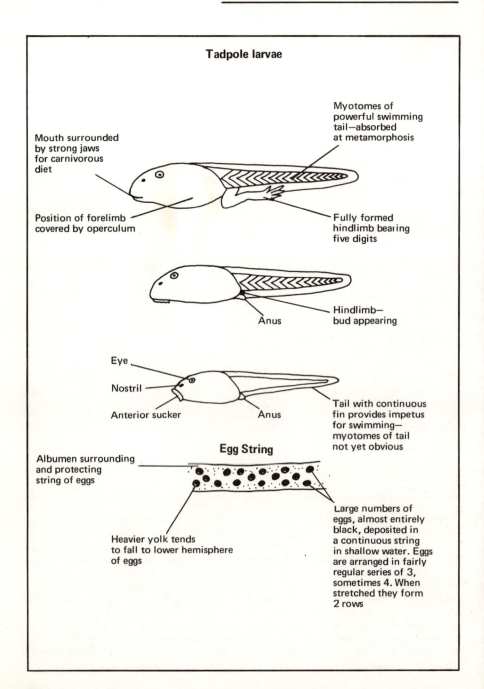

Tadpole larvae

Myotomes of powerful swimming tail—absorbed at metamorphosis

Mouth surrounded by strong jaws for carnivorous diet

Position of forelimb covered by operculum

Fully formed hindlimb bearing five digits

Anus

Hindlimb—bud appearing

Eye

Nostril

Anterior sucker

Anus

Tail with continuous fin provides impetus for swimming—myotomes of tail not yet obvious

Egg String

Albumen surrounding and protecting string of eggs

Heavier yolk tends to fall to lower hemisphere of eggs

Large numbers of eggs, almost entirely black, deposited in a continuous string in shallow water. Eggs are arranged in fairly regular series of 3, sometimes 4. When stretched they form 2 rows

PHYLUM CHORDATA	***Triturus***
SUBPHYLUM VERTEBRATA [CRANIATA]	Newt
CLASS AMPHIBIA	
ORDER URODELA [CAUDATA]	
GENUS *Triturus*	

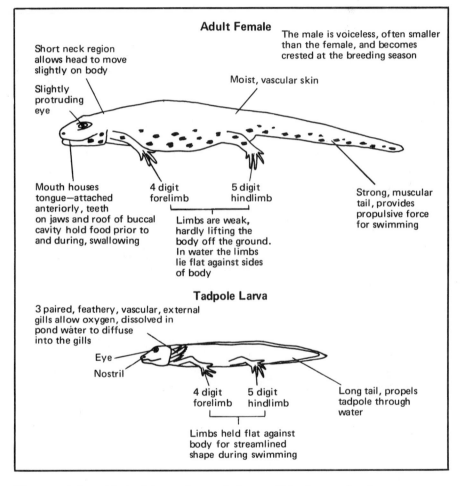

Adult Female

The male is voiceless, often smaller than the female, and becomes crested at the breeding season

Short neck region allows head to move slightly on body

Slightly protruding eye

Moist, vascular skin

Mouth houses tongue—attached anteriorly, teeth on jaws and roof of buccal cavity hold food prior to and during, swallowing

4 digit forelimb

5 digit hindlimb

Limbs are weak, hardly lifting the body off the ground. In water the limbs lie flat against sides of body

Strong, muscular tail, provides propulsive force for swimming

Tadpole Larva

3 paired, feathery, vascular, external gills allow oxygen, dissolved in pond water to diffuse into the gills

Eye

Nostril

4 digit forelimb

5 digit hindlimb

Long tail, propels tadpole through water

Limbs held flat against body for streamlined shape during swimming

The newt is found in moist grass in wooded areas. It feeds voraciously on worms, slugs and snails. In dry conditions newts curl up together to prevent water loss. In cold conditions they hibernate under stones.

In spring they emerge and return to the pond to feed, prior to the complex courtship during which the male deposits usually 1 spermatophore on the pond bottom. The female takes this into her cloaca and ejects the mucilaginous stump, then later deposits singly, fertilised, jelly-coated eggs onto water weeds or stones. A few weeks later, limbless tadpole larvae, with external gills, emerge and feed on invertebrates. Soon the forelimbs appear, the gills enlarge and the hindlimbs emerge. The gills function until metamorphosis when they disappear as the lungs are then fully formed, and tiny tailed newts leave the pond, but will return to breed in approximately three years time when they are mature.

Salamandra
Salamander

PHYLUM CHORDATA

SUBPHYLUM VERTEBRATA [CRANIATA]

CLASS AMPHIBIA

ORDER URODELA [CAUDATA]

GENUS *Salamandra*

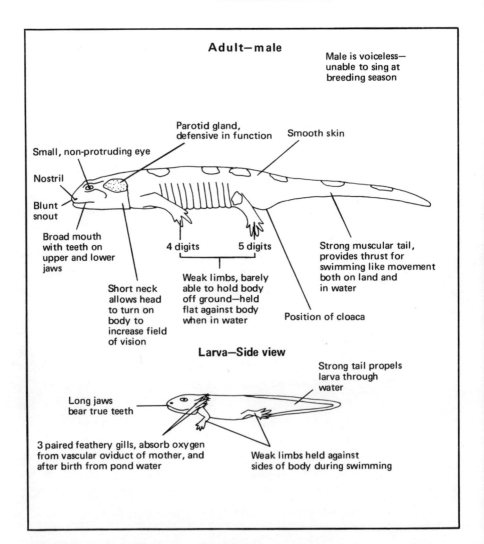

Adult—male

Male is voiceless—unable to sing at breeding season

Parotid gland, defensive in function

Smooth skin

Small, non-protruding eye

Nostril

Blunt snout

Broad mouth with teeth on upper and lower jaws

4 digits

5 digits

Short neck allows head to turn on body to increase field of vision

Weak limbs, barely able to hold body off ground—held flat against body when in water

Strong muscular tail, provides thrust for swimming like movement both on land and in water

Position of cloaca

Larva—Side view

Strong tail propels larva through water

Long jaws bear true teeth

3 paired feathery gills, absorb oxygen from vascular oviduct of mother, and after birth from pond water

Weak limbs held against sides of body during swimming

The salamander is terrestrial, usually solitary but several may shelter together in the same hole. It feeds on insects, worms, snails, slugs and beetles.

Mating occurs in July on land, when the male deposits spermatophores and the female positions her cloaca over a spermatophore, which enters and liberates the sperm, which are stored in spermothecae in the cloaca. Later the sperm migrate up the oviducts to fertilise the eggs which are retained in the oviducts. The female enters a pond and bears live 10—15 carnivorous, tadpole larvae with external gills, fore- and hindlimbs. At metamorphosis the gills degenerate, the lungs expand, and the tail and weak limbs are retained and the young salamander leaves the pond.

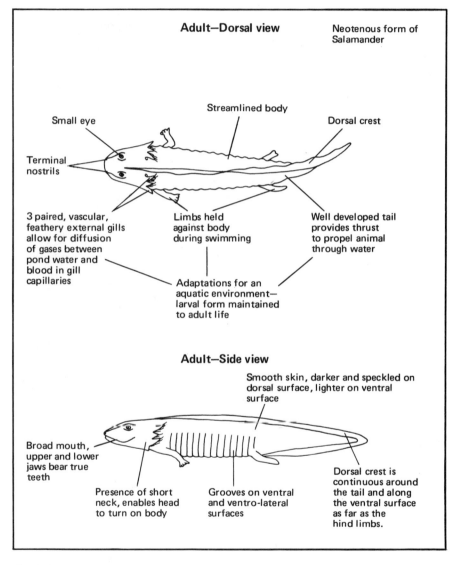

Adult—Dorsal view

Neotenous form of Salamander

Streamlined body

Small eye

Dorsal crest

Terminal nostrils

3 paired, vascular, feathery external gills allow for diffusion of gases between pond water and blood in gill capillaries

Limbs held against body during swimming

Well developed tail provides thrust to propel animal through water

Adaptations for an aquatic environment— larval form maintained to adult life

Adult—Side view

Smooth skin, darker and speckled on dorsal surface, lighter on ventral surface

Broad mouth, upper and lower jaws bear true teeth

Presence of short neck, enables head to turn on body

Grooves on ventral and ventro-lateral surfaces

Dorsal crest is continuous around the tail and along the ventral surface as far as the hind limbs.

Axolotls are aquatic and are found in the lakes of Colorado, Wyoming and Mexico. They are populations of species that fail to metamorphose for a terrestrial life. The adult size is attained, but the larval form is extended throughout life. The axolotl becomes sexually mature and breeds at the external gill stage. This neotenous form may be due to a low iodine intake and other factors; feeding on thyroid extract will induce metamorphosis.

Class Reptilia

Poikilothermic.

Terrestrial and aquatic—fresh water and marine.

External Features

Skin—dry bears horny scales or plates (scutes) which may overlap bony plates.

Limbs—two pairs, their rudiments or complete absence.

Visceral clefts—never develop gills either in embryo or adult.

Teeth—present in all reptiles except Testudines, usually uniform type.

Cloaca present.

Internal Features

Temporal vacuities present in all reptilian skulls except Testudines.

Ear—middle ear chamber usually present—has only one bone.

Lungs—respiratory organs. Breathe atmospheric air. Heart divided into two auricles and one ventricle, which is partly divided by a septum—except in crocodilians which have fully divided ventricle and thus four chambers.

Adult kidney metanephric—excretes uric acid or urea.

Fertilisation is internal.

Males usually possess either single or paired intromittent organs.

Eggs are cleidoic—large, yolky and enclosed in a leathery or calcareous shell and laid on land.

Embryo develops within embryonic membranes.

No larval stage. Some species are ovoviviparous.

REFERENCES FOR CLASS REPTILIA

Bellairs, A. d'A.	1957	Reptiles	London
Brongersma, L.D.	1967	British Turtles. Trustees of the British Museum (Natural History)	London
Carr, A.	1968	The Turtle	London
Hellmich, W.	1962	Reptiles and Amphibians of Europe	London
Smith, M.	1951	The British Amphibians and Reptiles	London

Suggested films for Class Reptilia

Rank Film Library	Life Story of the Snake	Colour	9 mins.
Encyclopaedia Britannica Films	Reptiles	Colour	14 mins.

ORDERS

TESTUDINES—turtles, terrapins and tortoises

1. Body short—enclosed by shell. Dorsal part—carapace. Ventral part —plastron. Each composed of inner plates of bone covered by outer plates or scutes of horny material or skin. Head, neck, tail, limbs, covered by scales, can be withdrawn or partly withdrawn inside shell.
2. Temporal vacuities absent from skull.
3. Teeth absent—horny bill or beak present.
4. Tongue not specialised.
5. Limbs of aquatic forms often modified to form paddles.
6. Tail usually short.
7. Single male intromittent organ.

e.g. *Chelonia*—green turtle

SQUAMATA—lizards, snakes

1. Body long—covered by horny scales.
2. Temporal vacuities present in skull— two, an upper and a lower one.
3. Teeth present—simple peg shape, but may show great specialisation.
4. Tongue specialised—thought to transfer scent particles to Jacobson's organ, olfactory and gustatory organ.
5. Limbs may be reduced or entirely absent.
6. Tail usually long.
7. Paired male intromittent organ.

ORDER SQUAMATA
Suborders
Lacertilia—lizards

1. One or more pairs of limbs may be absent.
2. Three movable eyelids including a nictitating membrane, although this may be reduced and eyelids may be fused.
3. Tympanum usually visible. Middle ear cavity present.
4. Numerous rows of scales on belly.
5. Tail often long.

e.g. *Lacerta*—common lizard

Ophidia—snakes

1. Limbs almost always entirely absent.
2. Movable eyelids absent. Eye covered by transparent spectacle.
3. Tympanum absent. Middle ear cavity absent.
4. Usually one row of scales on belly.
5. Body long.

e.g. *Natrix*—grass snake

(There is much grading one to another)

Chelonia

Green or edible turtle

PHYLUM CHORDATA	
SUBPHYLUM VERTEBRATA [CRANIATA]	
CLASS REPTILIA	
ORDER TESTUDINES	
GENUS *Chelonia*	

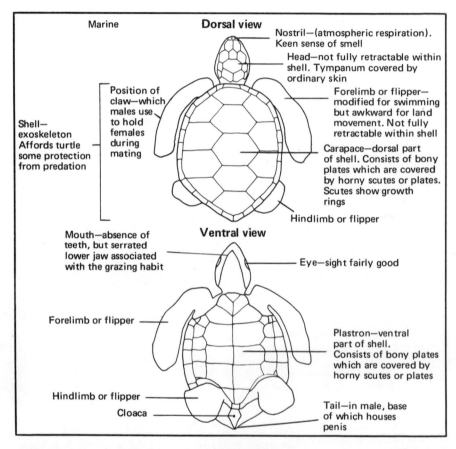

Marine

Dorsal view

Nostril—(atmospheric respiration). Keen sense of smell

Head—not fully retractable within shell. Tympanum covered by ordinary skin

Position of claw—which males use to hold females during mating

Forelimb or flipper—modified for swimming but awkward for land movement. Not fully retractable within shell

Shell—exoskeleton Affords turtle some protection from predation

Carapace—dorsal part of shell. Consists of bony plates which are covered by horny scutes or plates. Scutes show growth rings

Hindlimb or flipper

Ventral view

Mouth—absence of teeth, but serrated lower jaw associated with the grazing habit

Eye—sight fairly good

Forelimb or flipper

Plastron—ventral part of shell. Consists of bony plates which are covered by horny scutes or plates

Hindlimb or flipper

Cloaca

Tail—in male, base of which houses penis

The green turtle is normally an inhabitant of tropical and subtropical waters, but it has on occasions been found in temperate waters. It is a good swimmer but is cumbersome on land. The adult turtle feeds on sea grass, but is thought not to be totally herbivorous, since small crustaceans and molluscs have been found in the stomach.

Breeding usually occurs many hundreds of miles from the feeding grounds. Mating occurs in the water before or shortly after egg laying. The female goes onto the beach to lay her eggs, which are deposited in a nest excavated by her in the sand. The eggs are laid in a clutch of 100. The female may nest 3—5 times during the season at intervals of 12 days. The eggs are covered with sand where they remain for 60 days, after which time the young hatch out and emerge from the sand. They move en masse towards the sea. They swim immediately and during their young life are carnivorous—swooping on their prey.

The green turtle and its eggs are of economic importance in certain countries where they are a source of food.

Chelonia
Green or edible turtle

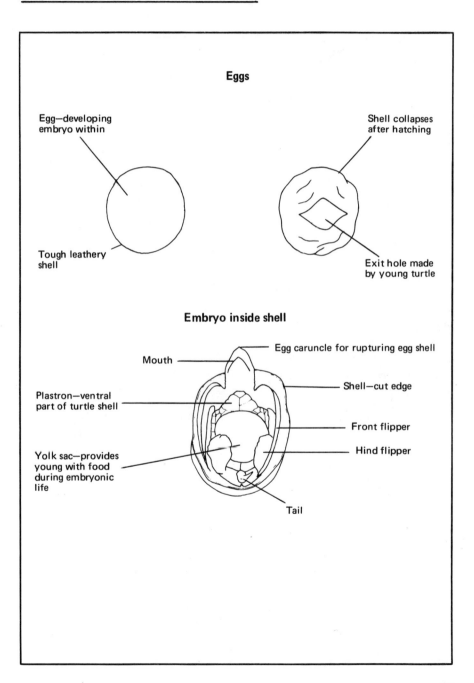

Eggs

Egg—developing embryo within

Tough leathery shell

Shell collapses after hatching

Exit hole made by young turtle

Embryo inside shell

Mouth

Plastron—ventral part of turtle shell

Yolk sac—provides young with food during embryonic life

Egg caruncle for rupturing egg shell

Shell—cut edge

Front flipper

Hind flipper

Tail

Lacerta
Common or viviparous lizard

PHYLUM CHORDATA	
SUBPHYLUM VERTEBRATA [CRANIATA]	
CLASS REPTILIA	
ORDER SQUAMATA	
GENUS *Lacerta*	

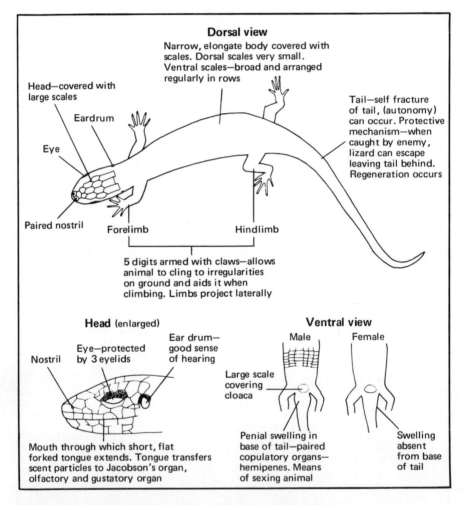

Dorsal view
Narrow, elongate body covered with scales. Dorsal scales very small. Ventral scales—broad and arranged regularly in rows

Head—covered with large scales

Eardrum

Eye

Paired nostril

Forelimb

Hindlimb

Tail—self fracture of tail, (autonomy) can occur. Protective mechanism—when caught by enemy, lizard can escape leaving tail behind. Regeneration occurs

5 digits armed with claws—allows animal to cling to irregularities on ground and aids it when climbing. Limbs project laterally

Head (enlarged)

Nostril

Eye—protected by 3 eyelids

Ear drum— good sense of hearing

Mouth through which short, flat forked tongue extends. Tongue transfers scent particles to Jacobson's organ, olfactory and gustatory organ

Ventral view

Male

Female

Large scale covering cloaca

Penial swelling in base of tail—paired copulatory organs— hemipenes. Means of sexing animal

Swelling absent from base of tail

The common lizard is found in almost all sunny areas in the British Isles. It can swim and sometimes climbs. They tend to live in colonies before the mating season, soon after this they disperse. Their food consists of insects and spiders; the former are often sought in water. Food is swallowed whole, but if too large is chewed first.

Mating occurs in April and May. Fertilisation is internal. Most lizards lay eggs with tough parchment shells, where the embryo will develop outside the mother's body—oviparous. The common lizard retains the eggs within her body for 3 months and the young are born alive, enclosed only by a very thin membrane which they puncture with their small egg tooth shortly after birth—ovoviviparous

Hibernation occurs between October and April in Britain.

PHYLUM CHORDATA	
SUBPHYLUM VERTEBRATA [CRANIATA]	
CLASS REPTILIA	
ORDER SQUAMATA	
GENUS *Lacerta*	

Lacerta
Common or viviparous lizard

Eggs—removed from female lizards

5–8 eggs produced in an average litter

The eggs containing young lizards are laid in a cavity excavated by the mother which is then concealed

Young lizard—emerging from shell

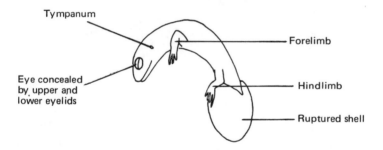

Tympanum

Eye concealed by upper and lower eyelids

Forelimb

Hindlimb

Ruptured shell

Section through skin of Lacerta

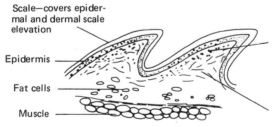

Scale—covers epidermal and dermal scale elevation

Epidermis

Fat cells

Muscle

Pigment cell—responsible for colour of skin and may be involved in the process of colour change

Dermis—consists of connective tissue nerves, blood vessels and pigment cells

Reptilian integument is fairly waterproof. The skin has few glands and when present they are local in distribution

The young lizard emerges from the shell shortly after egg laying or whilst still in the oviduct. They are fully mobile about an hour after hatching or birth and feed after a few hours.

Natrix
Grass snake or ringed snake

PHYLUM CHORDATA

SUBPHYLUM VERTEBRATA [CRANIATA]

CLASS REPTILIA

ORDER SQUAMATA

GENUS *Natrix*

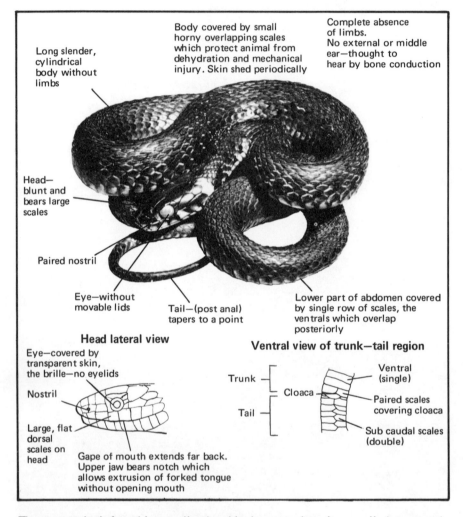

Long slender, cylindrical body without limbs

Body covered by small horny overlapping scales which protect animal from dehydration and mechanical injury. Skin shed periodically

Complete absence of limbs.
No external or middle ear—thought to hear by bone conduction

Head—blunt and bears large scales

Paired nostril

Eye—without movable lids

Tail—(post anal) tapers to a point

Lower part of abdomen covered by single row of scales, the ventrals which overlap posteriorly

Head lateral view

Eye—covered by transparent skin, the brille—no eyelids

Nostril

Large, flat dorsal scales on head

Gape of mouth extends far back. Upper jaw bears notch which allows extrusion of forked tongue without opening mouth

Ventral view of trunk—tail region

Trunk

Cloaca

Tail

Ventral (single)

Paired scales covering cloaca

Sub caudal scales (double)

The grass snake is found in woodland and hedgerows where it may climb trees and marshy ground where it swims. Its diet consists of frogs, newts, fish and also small birds and mammals which it swallows whole.

Mating occurs in the spring and fertilisation is internal. The oval, flexible eggs are laid, 2 months after mating in soft masses of 30—40 in manure heaps, rotting leaves etc. which provide the necessary warmth for their development. Hatching occurs after six to ten weeks. The young snake ruptures the shell by means of the egg tooth which is later shed. The young snakes remain in hiding for some time as they are prey to some animals.

Hibernation occurs between October and April, in Britain.

Natrix
Grass snake or ringed snake

Eggs—stick together after laying. Female may stay near eggs and show aggresive behaviour if disturbed

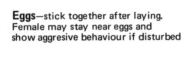

Soft membranous shell

Egg—cut open to show developing embryo

Stump of yolk sac attached to belly of embryo

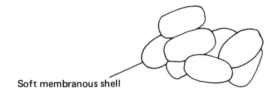

Deflected shell. Shell starts to shrivel prior to hatching

Mouth

Young snake—recently hatched (scales not shown)

Eye

Nostril

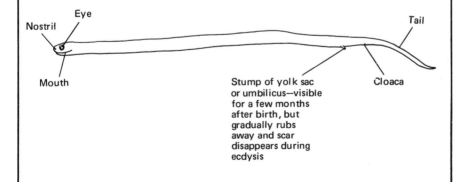

Tail

Mouth

Stump of yolk sac or umbilicus—visible for a few months after birth, but gradually rubs away and scar disappears during ecdysis

Cloaca

Class Aves

Homeothermic—usually capable of flight.
Terrestrial and aerial.

External Features

Skin covered by feathers.
Three digit forelimbs form wings for flight.
Hindlimbs strong for walking and running—feet
covered by horny scales.
Jaws toothless, covered by horny beak.

Internal Features

Pectoral girdle strong, sternum large for attachment of
muscles moving wings.
Pelvic girdle large, fused to several sacral vertebrae for
strength for bipedal gait.
Ear consists of inner, middle and outer chambers—
lacks pinna.
Air sacs present in strong light bones.
Lungs are the sole respiratory surface—visceral clefts
never bear gills.
Heart consists of two auricles and two ventricles.
Adult excretory organs are metanephric kidneys
which excrete uric acid.
Three chambered cloaca present—receives ureters,
genital ducts and rectum.

Fertilisation internal.
Large yolked, cleidoic eggs laid in nests.
Embryo develops within embryonic membranes.
No larval stage.
Parental care before and after hatching from the egg.

ORDER COLUMBIFORMES

Birds capable of rapid flight.
Grain eaters with short rectal caeca.
e.g. *Columba*—pigeon

REFERENCE FOR CLASS AVES
Gilliard, E.T.   1959   Living Birds of the World   London

Columba
Pigeon

External features—side view

Lead coloured bill

Large eyes—
vision acute

Contour feathers
cover head and neck

Dense plumage

Fusiform body offers little
resistance to air during
flight.
Coat of feathers provides
light warm covering and
restricts heat loss from
body surface. The feathers
are shed at intervals and
replaced

Red scaly
feet of the
hindlimbs
which are
modified to
support the
weight of
the whole
body—
bipedal gait

Remiges—quill
feathers of the
wings, which are
the forelimbs
modified for
flight

Retrices—quill feathers
of the short tail, are
used for steering, braking
and maintaining stability
during flight.
The uropygial gland
present on the tail
is the sole skin gland
and secretes an
oily substance (used in
some birds for preening
but not in pigeons).

Domestic pigeons are the descendants of the wild rock dove which is found along
the rocky coasts of western and southern Europe. Their flight is swift, at speeds
between 30 to 60 miles per hour, often low over water or land. They also glide
freely about cliff faces.

Columba
Pigeon

PHYLUM CHORDATA

SUBPHYLUM VERTEBRATA [CRANIATA]

CLASS AVES

ORDER COLUMBIFORMES

GENUS *Columba*

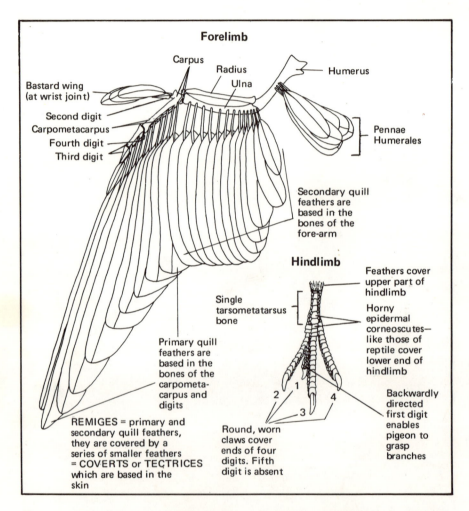

Forelimb

Carpus

Radius

Ulna

Humerus

Bastard wing (at wrist joint)

Second digit

Carpometacarpus

Fourth digit

Third digit

Pennae Humerales

Secondary quill feathers are based in the bones of the fore-arm

Hindlimb

Feathers cover upper part of hindlimb

Horny epidermal corneoscutes— like those of reptile cover lower end of hindlimb

Single tarsometatarsus bone

Primary quill feathers are based in the bones of the carpometacarpus and digits

2 1

3

4

Round, worn claws cover ends of four digits. Fifth digit is absent

Backwardly directed first digit enables pigeon to grasp branches

REMIGES = primary and secondary quill feathers, they are covered by a series of smaller feathers = COVERTS or TECTRICES which are based in the skin

The nest is a slight structure built of small twigs, roots and dead stems of grasses. According to species, it may be placed on a horizontal branch of a tree, on a ledge in a cave, a crevice in a cliff or equivalent sites on buildings. Both sexes usually assist in its construction and in incubating the two white eggs which are laid in spring or summer. Incubation lasts for 14—19 days; then the young hatch as blind, featherless nestlings and remain in the nest about a month. They are brooded by both parents and fed on "pigeon's milk" which is a regurgitated secretion of the crop lining and it is injected into the nestling when it places its bill into that of the parent. Later they are fed on softened grain. Sparse hairy down and then feathers develop enabling them to fly and leave the nest. Two or three broods of young are reared annually.

Columba
Pigeon

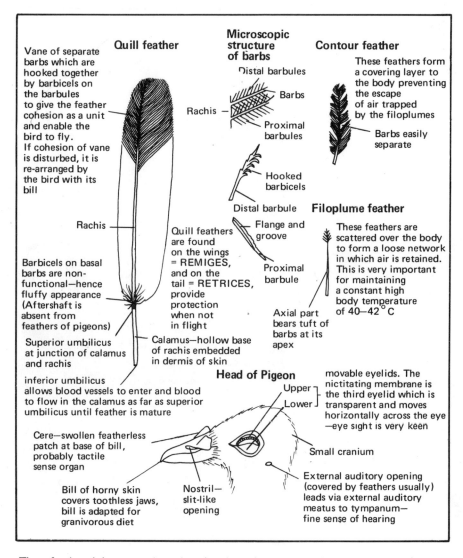

Quill feather

Vane of separate barbs which are hooked together by barbicels on the barbules to give the feather cohesion as a unit and enable the bird to fly. If cohesion of vane is disturbed, it is re-arranged by the bird with its bill

Rachis

Barbicels on basal barbs are non-functional—hence fluffy appearance (Aftershaft is absent from feathers of pigeons)

Superior umbilicus at junction of calamus and rachis

inferior umbilicus allows blood vessels to enter and blood to flow in the calamus as far as superior umbilicus until feather is mature

Quill feathers are found on the wings = REMIGES, and on the tail = RETRICES, provide protection when not in flight

Calamus—hollow base of rachis embedded in dermis of skin

Microscopic structure of barbs

Distal barbules

Barbs

Rachis

Proximal barbules

Hooked barbicels

Distal barbule

Flange and groove

Proximal barbule

Axial part bears tuft of barbs at its apex

Contour feather

These feathers form a covering layer to the body preventing the escape of air trapped by the filoplumes

Barbs easily separate

Filoplume feather

These feathers are scattered over the body to form a loose network in which air is retained. This is very important for maintaining a constant high body temperature of 40–42°C

Head of Pigeon

Upper ⎤
Lower ⎦

movable eyelids. The nictitating membrane is the third eyelid which is transparent and moves horizontally across the eye —eye sight is very keen

Small cranium

Cere—swollen featherless patch at base of bill, probably tactile sense organ

Bill of horny skin covers toothless jaws, bill is adapted for granivorous diet

Nostril— slit-like opening

External auditory opening (covered by feathers usually) leads via external auditory meatus to tympanum— fine sense of hearing

They feed mainly on seeds and grains, but also on peas, beans, potatoes, insects and snails.

Pigeons are monogamous—they maintain the same mate for life. Many pigeons are gregarious and often fly, feed, rest or roost in company. Flocking facilitates utilisation of scattered food supplies, pairing; it also enables the young to benefit from the knowledge and experience of the older birds and it also acts as a check on possible predators.

Class Mammalia

Homeothermic.
Terrestrial, fossorial, aerial, arboreal and aquatic.

External Features

Skin produces hair, sweat glands and sebaceous glands.
Milk producing glands—modified skin glands secrete
milk for nourishment of young.
External ear—pinna.
Teeth—heterodont, rooted in sockets in jaws.
Pentadactyl limbs—2 pairs, show much specialisation.

Internal Features

Lower jaw or mandible consists of a single bone, the
dentary.
Brain—well developed.
Middle ear—three bones or ossicles present in middle
ear chamber.
Heart—complete division between right and left halves
—two auricles, two ventricles.
Lungs with alveoli.
Epiglottis separates trachea from oesophagus.
Diaphragm—muscular sheet separates thoracic and
abdominal body cavities.
Kidneys—metanephric—excrete urea.

Fertilisation is internal.
Males possess an intromittent organ.
Eggs usually small with little yolk.
Embryo develops within embryonic membranes. No
larval stage. Few young produced at any one time.
Suckled by mother. Parental care during and
frequently after lactation.

Subclass Theria

Viviparous—young born alive.
Anus and urino-genital apertures distinct—no cloaca.

Infraclasses

METATHERIA

1. Marsupial mammals.
2. Young born at early stage in development.

3. Nipples covered by pouch in which young continues development.
4. Suckling period long and suckling irregular.
5. Monophyodont—usually one set of teeth.
6. Double uteri in female. Forked penis in male.
7. Small brain—low intelligence.

    e.g. **Order Marsupialia**—kangaroo

EUTHERIA

1. Placental mammals.
2. Young born at later stage in development. Nourished by placenta in uterus.
3. Pouch absent. Nipples exposed.

4. Suckling period short. Suckling regular.
5. Diphyodont—two sets of teeth, milk and permanent.
6. Single uterus in female. Penis not forked in male.
7. Larger brain—wider range of intelligence.

    e.g. **Order Insectivora**—mole
        Order Chiroptera—bat
        Order Primates—chimpanzee
        Order Lagomorpha—rabbit
        Order Rodentia—rat
        Order Cetacea—whale
        Order Carnivora—cat
        Order Pinnipedia—seal
        Order Perissodactyla—horse
        Order Artiodactyla—sheep

REFERENCES FOR CLASS MAMMALIA

Aller, G.M.	1962	Bats	London
Bourlière, F.	1959	Mammals of the World	London
Corbet, G.B.	1964	The Identification of British Mammals Trustees of the British Museum (Natural History)	London
Darling, L.	1961	Kangaroos and other animals with pouches	London
Fraser, F.C.	1949	British Whales, Dolphins and Porpoises Trustees of the British Museum (Natural History)	London
Harrison, R.J. and King, J.E.	1965	Marine Mammals	London

Hewer, H.R.	1962	Animals of Britain (Grey Seal)	London
Life Nature Library	1965	The Mammals	Netherlands
Matthews, L.H.	1952	British Mammals	London
Napier, J.R. and Napier, P.H.	1967	A Handbook of Living Primates	London
Rowett, H.G.O.	1957	The Rat as a Small Mammal	London
Sanderson, I.I.	1959	Living Mammals of the World	London
Schultz, A.H.	1969	A Life of Primates	London
Sisson, S.	1964	Anatomy of Domestic Animals	Philadelphia
Slijper, E.J.	1962	Whales	London
Southern, H.N.	1964	The Handbook of British Mammals	London
Street, P.	1961	Mammals of the British Isles	London
Thompson, H.V. and Worden, A.N.	1956	The Rabbit	London
Yerkes, R.M. and Yerkes, A.W.	1929	The Great Apes	U.S.A.
Young, J.Z.	1957	Life of Mammals	London

Suggested films for Class Mammalia

Bolton, Hawker Films Ltd.	The Birth of a Red Kangaroo	Colour	21 mins.

PHYLUM CHORDATA	**_Macropus_**
SUBPHYLUM VERTEBRATA [CRANIATA]	Kangaroo
CLASS MAMMALIA	
INFRACLASS METATHERIA	
ORDER MARSUPIALIA	
GENUS _Macropus_	

External features of Male Kangaroo

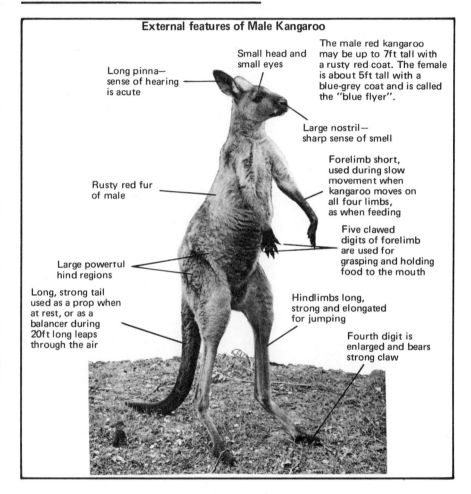

Small head and small eyes

The male red kangaroo may be up to 7ft tall with a rusty red coat. The female is about 5ft tall with a blue-grey coat and is called the "blue flyer".

Long pinna— sense of hearing is acute

Large nostril— sharp sense of smell

Forelimb short, used during slow movement when kangaroo moves on all four limbs, as when feeding

Rusty red fur of male

Five clawed digits of forelimb are used for grasping and holding food to the mouth

Large powerful hind regions

Long, strong tail used as a prop when at rest, or as a balancer during 20ft long leaps through the air

Hindlimbs long, strong and elongated for jumping

Fourth digit is enlarged and bears strong claw

Kangaroos are abundant in the broad dry interior plains and highlands of Australia and also in Papua and the Aru islands. They are herbivores and graze on the hills and plains in the early morning and late afternoons. During the heat of the day they rest in the shade of trees and rocky outcrops or they make shallow pits in the earth.

Kangaroos are gregarious; they travel in "mobs" ruled by a seven foot tall, grown, male— the "old man" or "boomer"—who defends his mob and drives all other grown males away. These young bucks spend a lonely wandering life until they are four years old and at maximum strength, when they challenge the "old man"—they lash out with their forelimbs, bite the neck veins and try to rip open the rival's abdomen with the large hoof of the hindlimb. If the younger kangaroo wins he becomes the new "old man".

Macropus
Kangaroo

PHYLUM CHORDATA

SUBPHYLUM VERTEBRATA [CRANIATA]

CLASS MAMMALIA

INFRACLASS METATHERIA

ORDER MARSUPIALIA

GENUS *Macropus*

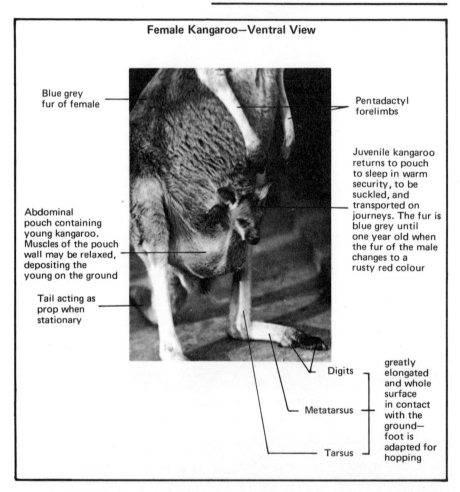

Female Kangaroo—Ventral View

Blue grey fur of female

Pentadactyl forelimbs

Juvenile kangaroo returns to pouch to sleep in warm security, to be suckled, and transported on journeys. The fur is blue grey until one year old when the fur of the male changes to a rusty red colour

Abdominal pouch containing young kangaroo. Muscles of the pouch wall may be relaxed, depositing the young on the ground

Tail acting as prop when stationary

Digits

Metatarsus

Tarsus

greatly elongated and whole surface in contact with the ground— foot is adapted for hopping

A young female is ready to mate at only 18 months of age. Breeding occurs throughout the year. 33 days after fertile mating the female begins to lick clean the inside of the pouch and adopts the birth position with her back supported against a tree. The young is then born and immediately crawls up the belly fur into the puch. It attaches itself, without the mother's aid, to 1 of 4 nipples and is suckled on milk pumped into its mouth by the mammary gland. Mating may occur again immediately after birth, but the embryo remains dormant in the uterus and birth will not occur until the first young has permanently left the pouch. The second young then enters the pouch and is suckled on another nipple, this may be followed by mating, so that the kangaroo may have, at one time, a dependent young-at-foot, a young in the pouch and a dormant embryo in the uterus.

40

PHYLUM CHORDATA

SUBPHYLUM VERTEBRATA [CRANIATA]

CLASS MAMMALIA

INFRACLASS METATHERIA

ORDER MARSUPIALIA

GENUS *Macropus*

Macropus
Kangaroo

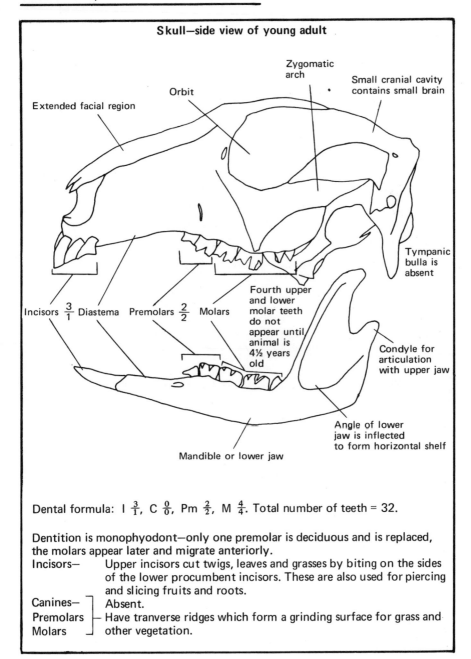

Skull—side view of young adult

Zygomatic arch

Orbit

Small cranial cavity contains small brain

Extended facial region

Tympanic bulla is absent

Incisors $\frac{3}{1}$ Diastema Premolars $\frac{2}{2}$ Molars

Fourth upper and lower molar teeth do not appear until animal is 4½ years old

Condyle for articulation with upper jaw

Angle of lower jaw is inflected to form horizontal shelf

Mandible or lower jaw

Dental formula: I $\frac{3}{1}$, C $\frac{0}{0}$, Pm $\frac{2}{2}$, M $\frac{4}{4}$. Total number of teeth = 32.

Dentition is monophyodont—only one premolar is deciduous and is replaced, the molars appear later and migrate anteriorly.

Incisors— Upper incisors cut twigs, leaves and grasses by biting on the sides of the lower procumbent incisors. These are also used for piercing and slicing fruits and roots.

Canines— ⎤ Absent.
Premolars ⎬ Have tranverse ridges which form a grinding surface for grass and
Molars ⎦ other vegetation.

Macropus
Kangaroo

PHYLUM CHORDATA

SUBPHYLUM VERTEBRATA [CRANIATA]

CLASS MAMMALIA

INFRACLASS METATHERIA

ORDER MARSUPIALIA

GENUS *Macropus*

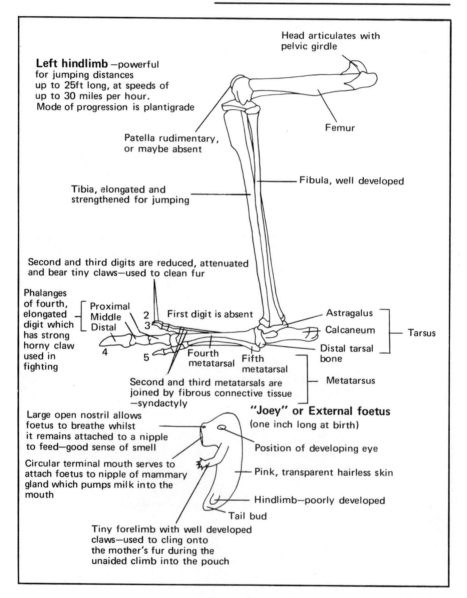

Left hindlimb —powerful for jumping distances up to 25ft long, at speeds of up to 30 miles per hour. Mode of progression is plantigrade

Head articulates with pelvic girdle

Femur

Patella rudimentary, or maybe absent

Fibula, well developed

Tibia, elongated and strengthened for jumping

Second and third digits are reduced, attenuated and bear tiny claws—used to clean fur

Phalanges of fourth, elongated digit which has strong horny claw used in fighting

Proximal
Middle
Distal

2
3

First digit is absent

Astragalus

Calcaneum

Tarsus

Distal tarsal bone

4
5

Fourth metatarsal

Fifth metatarsal

Metatarsus

Second and third metatarsals are joined by fibrous connective tissue —syndactyly

Large open nostril allows foetus to breathe whilst it remains attached to a nipple to feed—good sense of smell

Circular terminal mouth serves to attach foetus to nipple of mammary gland which pumps milk into the mouth

"Joey" or External foetus
(one inch long at birth)

Position of developing eye

Pink, transparent hairless skin

Hindlimb—poorly developed

Tail bud

Tiny forelimb with well developed claws—used to cling onto the mother's fur during the unaided climb into the pouch

The kangaroo embryo is not attached to the uterus by a placenta. The foetus is therefore born at a very early stage in development and continues to develop externally in the pouch for about 235 days, then the young leaves the pouch. About 365 days after birth it is fully weaned and feeds entirely on young shoots.

Infraclass Eutheria

ORDER	INSECTIVORA	CHIROPTERA	PRIMATES
Habit	Fossorial, terrestrial arboreal and aquatic.	Aerial.	Arboreal and terrestrial.
Diet	Mainly insectivorous.	Insectivorous or frugivorous.	Herbivorous or omnivorous.
Teeth	Numerous, small cusped teeth.	Numerous cusped teeth. Insectivorous—cusped molars. Frugivorous—flat molar teeth. Upper incisors separated by a deep concavity or borne on a bone projecting from such a concavity.	Teeth not highly modified. Molars—bear blunt cusps.
Skull	Zygomatic arch may be incomplete. Small cranium. Long facial region.	Zygomatic arch—slender. Lower jaw—slender.	Large cranium—reduced facial region. Orbits—tend to face forwards. Foramen magnum faces downwards.
Gut	Stomach—simple. Intestine—simple. Caecum—absent.	Stomach: Insectivores—small, simple, flask shape. Frugivores—larger and more complex. Intestine: simple, lacks appendix.	Stomach—simple and capacious. Appendix present.
Limbs	Forelimbs—plantigrade. Hind limbs—plantigrade	Forelimbs modified to form wings. Hindlimbs short—knees turn outwards. Digits prehensile.	Forelimbs modified for manipulating food and climbing. Hindlimbs—plantigrade.
Example	*Talpa*—mole	*Pipistrellus*—pipistrelle bat	*Pan*—chimpanzee

ORDER	LAGOMORPHA	RODENTIA	CETACEA
Habit	Terrestrial and fossorial.	Terrestrial, arboreal and fossorial.	Marine.
Diet	Herbivorous.	Herbivorous or omnivorous.	Carnivorous or plankton feeder.
Teeth	Two pairs of upper incisors. 2nd pair small and immediately behind first pair. Incisors covered by enamel. Canines absent—diastema present. Molars when biting—lower jaw teeth inside upper jaw teeth.	One pair of upper incisors. Incisors—chisel shaped with persistent pulp. Enamel on front only. Canines absent—diastema present. Molars when biting—upper jaw teeth inside lower jaw teeth.	Carnivores—simple conical teeth—homodont. Plankton feeders—whale bone or baleen, row of triangular hairy plates on each side of palate.
Skull	Zygomatic arch—large and complete. Small cranium.	Zygomatic arch—large and complete.	Large skull. Telescoping of bones in skull. Long jaws. Skull of toothed whales—asymmetrical.
Gut	Stomach—simple. Intestine—long. Caecum—large. Appendix—large. Caecotrophy—food passes twice through alimentary canal. Soft night pellets eaten for bacterially produced metabolites.	Stomach—simple. Intestine—long with fairly voluminous caecum. Appendix short.	Stomach—three chambered. First chamber often containing stones—grinds up food. Intestine long. No clearly defined caecum.
Limbs	Forelimbs—plantigrade, modified for burrowing. Hindlimbs—plantigrade, modified for leaping.	Forelimbs—plantigrade, often used for carrying food. Hindlimbs—plantigrade, certain species known to exhibit bipedalism.	Forelimbs—modified to form flipper. Hindlimbs—absent.
Example	*Oryctolagus*—rabbit	*Rattus*—black rat	*Globicephala*—pilot whale

ORDER	CARNIVORA	PINNIPEDIA	PERISSODACTYLA	ARTIODACTYLA
Habit	Arboreal and terrestrial.	Aquatic and terrestrial.	Terrestrial.	Terrestrial.
Diet	Carnivorous.	Carnivorous.	Herbivorous.	Herbivorous.
Teeth	Canines— strong and deep rooted. Carnassial teeth —for cracking bones and shearing flesh.	Canines—large. Carnassial teeth absent. Cheek teeth very much identical— conical and laterally compressed. Number of cusps varies.	Incisors—used for cropping. Canines reduced or absent, often diastema. Premolars and molars similar— modified to form one continuous grinding surface.	Upper incisors—may be absent. Crop with gums. Canines—sometimes modified to form tusks. Premolars and molars dissimilar. Premolars simpler than molars.
Skull	Zygomatic arch strong. Large cranium.	Zygomatic arch —strong. Large orbits. Short snout.	Cranium— small. Large, elongate facial region.	Large facial region. Horns may grow from skull. Some shed annually— antlers. Others never shed.
Gut	Stomach— simple. Short alimentary canal. Caecum small or absent.	Stomach— simple. Intestine—long. Caecum absent or indistinct.	Stomach— simple. Large intestine and caecum— well developed for bacterial digestion.	Stomach— complex, ruminant type —four chambers —for bacterial digestion.
Limbs	Forelimbs— digitigrade. Hindlimbs— digitigrade. Claws sometimes retractile. Large forms— plantigrade.	Forelimbs— modified to form flippers. Hindlimbs— modified to form flippers.	Forelimbs— unguligrade. Reduction in digits except third. Hindlimbs— unguligrade.	Forelimbs— unguligrade. Reduction in digits except third and fourth. Hindlimbs— unguligrade.
Example	*Felis*—cat	*Halichoerus*— grey seal	*Equus*—horse	*Ovis*—sheep

Talpa
Mole

PHYLUM CHORDATA

SUBPHYLUM VERTEBRATA [CRANIATA]

CLASS MAMMALIA

INFRACLASS EUTHERIA

ORDER INSECTIVORA

GENUS *Talpa*

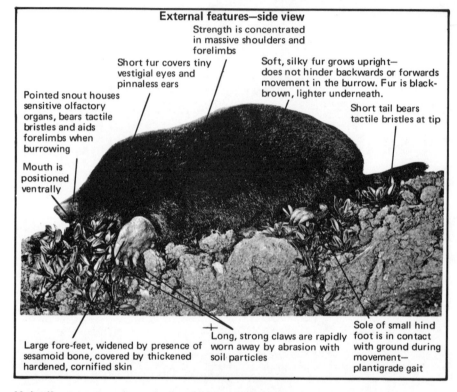

External features—side view

Strength is concentrated in massive shoulders and forelimbs

Short fur covers tiny vestigial eyes and pinnaless ears

Pointed snout houses sensitive olfactory organs, bears tactile bristles and aids forelimbs when burrowing

Mouth is positioned ventrally

Soft, silky fur grows upright—does not hinder backwards or forwards movement in the burrow. Fur is black-brown, lighter underneath.

Short tail bears tactile bristles at tip

Large fore-feet, widened by presence of sesamoid bone, covered by thickened hardened, cornified skin

Long, strong claws are rapidly worn away by abrasion with soil particles

Sole of small hind foot is in contact with ground during movement—plantigrade gait

Moles live permanently underground in tunnel systems in the soft soil of gardens, farmlands and deciduous woodlands throughout Europe.

They are intensely active by day and night. They have prodigious appetites. Within 24 hours they consume worms, grubs, slugs and small soil animals which aggregate to their own weight. Moles catch worms, bite the anterior end to paralyse them and then store the worms alive until conditions become adverse. Accordingly the moles are able to remain active during the winter.

The spade-like forelimbs act as scrapers when digging the tunnel system and also as levers for pushing the body through the soil which is then swept aside by the head and snout. In hard earth the fore-feet are brought in front of the snout to scrape away the earth, which is then swept aside by head movements or thrust up a shaft onto the surface to form a mole hill.

Moles are fiercely unsociable. In spring the mole captures a female and copulation occurs within the tunnel system. After a gestation period of 40 days about 4 naked, blind young are born in the leaf-lined nest, which is built at the cross roads of the tunnel system. They are suckled for three weeks. Life expectancy is about 3 years but moles are heavily preyed on by owls, foxes, stoats and badgers.

The fur of mole is of value for fashion trimmings.

PHYLUM CHORDATA

SUBPHYLUM VERTEBRATA [CRANIATA]

CLASS MAMMALIA

INFRACLASS EUTHERIA

ORDER INSECTIVORA

GENUS *Talpa*

Talpa
Mole

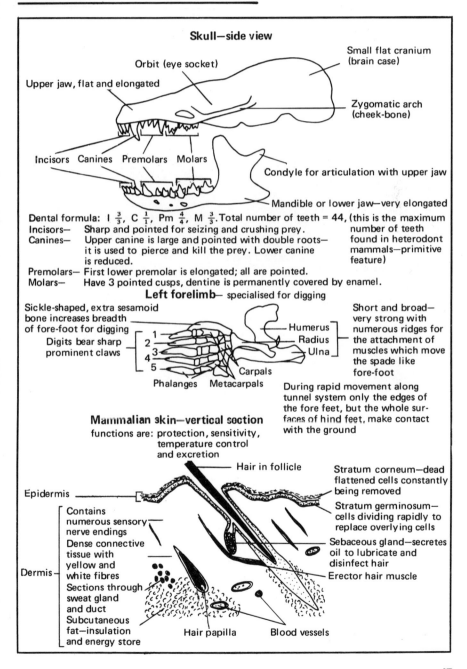

Skull—side view

Small flat cranium (brain case)

Orbit (eye socket)

Upper jaw, flat and elongated

Zygomatic arch (cheek-bone)

Incisors Canines Premolars Molars

Condyle for articulation with upper jaw

Mandible or lower jaw—very elongated

Dental formula: $I\frac{3}{3}$, $C\frac{1}{1}$, $Pm\frac{4}{4}$, $M\frac{3}{3}$. Total number of teeth = 44, (this is the maximum number of teeth found in heterodont mammals—primitive feature)

Incisors— Sharp and pointed for seizing and crushing prey.

Canines— Upper canine is large and pointed with double roots— it is used to pierce and kill the prey. Lower canine is reduced.

Premolars— First lower premolar is elongated; all are pointed.

Molars— Have 3 pointed cusps, dentine is permanently covered by enamel.

Left forelimb— specialised for digging

Sickle-shaped, extra sesamoid bone increases breadth of fore-foot for digging

Digits bear sharp prominent claws

1
2
3
4
5

Humerus
Radius
Ulna

Short and broad— very strong with numerous ridges for the attachment of muscles which move the spade like fore-foot

Carpals

Phalanges Metacarpals

During rapid movement along tunnel system only the edges of the fore feet, but the whole sur- faces of hind feet, make contact with the ground

Mammalian skin—vertical section

functions are: protection, sensitivity, temperature control and excretion

Hair in follicle

Stratum corneum—dead flattened cells constantly being removed

Epidermis

Stratum germinosum— cells dividing rapidly to replace overlying cells

Dermis—
Contains numerous sensory nerve endings
Dense connective tissue with yellow and white fibres
Sections through sweat gland and duct
Subcutaneous fat—insulation and energy store

Sebaceous gland—secretes oil to lubricate and disinfect hair

Erector hair muscle

Hair papilla Blood vessels

47

Pipistrellus
Pipistrelle bat

PHYLUM CHORDATA

SUBPHYLUM VERTEBRATA [CRANIATA]

CLASS MAMMALIA

INFRACLASS EUTHERIA

ORDER CHIROPTERA

GENUS *Pipistrellus*

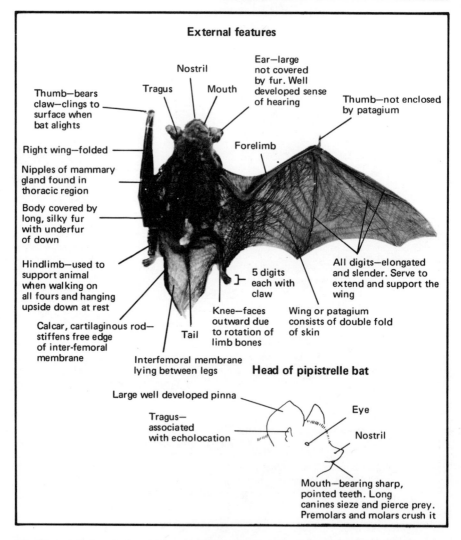

External features

Thumb—bears claw—clings to surface when bat alights

Right wing—folded

Nipples of mammary gland found in thoracic region

Body covered by long, silky fur with underfur of down

Hindlimb—used to support animal when walking on all fours and hanging upside down at rest

Calcar, cartilaginous rod—stiffens free edge of inter-femoral membrane

Tail

Interfemoral membrane lying between legs

Nostril

Tragus   Mouth

Ear—large not covered by fur. Well developed sense of hearing

Thumb—not enclosed by patagium

Forelimb

5 digits each with claw

Knee—faces outward due to rotation of limb bones

Wing or patagium consists of double fold of skin

All digits—elongated and slender. Serve to extend and support the wing

Head of pipistrelle bat

Large well developed pinna

Tragus—associated with echolocation

Eye

Nostril

Mouth—bearing sharp, pointed teeth. Long canines sieze and pierce prey. Premolars and molars crush it

The Pipistrelle bat is the most common bat in Britain. It has the ability to fly, but it does not reach very great heights. It is a gregarious animal and is found in colonies consisting of large numbers of individuals. The colonies are found in houses and trees.

They are nocturnal animals, coming out to feed at dusk. But during the day they may be found resting, head downward, hanging by the digits of hind limbs, with the wings folded.

Feeding is performed on the wing. Their diet is composed of insects. The

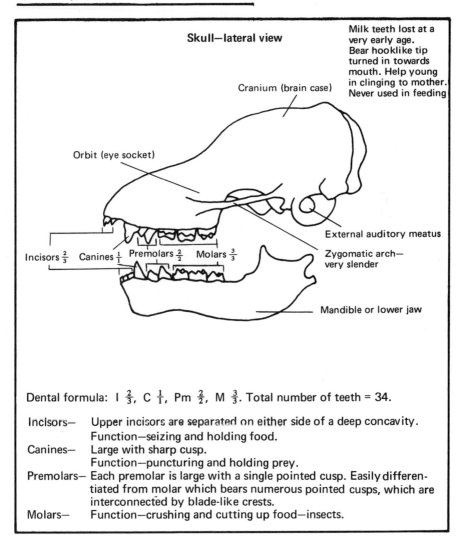

PHYLUM CHORDATA	***Pipistrellus***
SUBPHYLUM VERTEBRATA [CRANIATA]	Pipistrelle bat
CLASS MAMMALIA	
INFRACLASS EUTHERIA	
ORDER CHIROPTERA	
GENUS *Pipistrellus*	

Skull—lateral view

Milk teeth lost at a very early age. Bear hooklike tip turned in towards mouth. Help young in clinging to mother. Never used in feeding

Cranium (brain case)

Orbit (eye socket)

External auditory meatus

Incisors $\frac{2}{3}$ Canines $\frac{1}{1}$ Premolars $\frac{2}{2}$ Molars $\frac{3}{3}$

Zygomatic arch— very slender

Mandible or lower jaw

Dental formula: I $\frac{2}{3}$, C $\frac{1}{1}$, Pm $\frac{2}{2}$, M $\frac{3}{3}$. Total number of teeth = 34.

Incisors— Upper incisors are separated on either side of a deep concavity. Function—seizing and holding food.

Canines— Large with sharp cusp. Function—puncturing and holding prey.

Premolars— Each premolar is large with a single pointed cusp. Easily differentiated from molar which bears numerous pointed cusps, which are interconnected by blade-like crests.

Molars— Function—crushing and cutting up food—insects.

small insects are eaten during flight, but the large ones are immobilised by removal of head and limbs and kept in the upturned interfemoral membrane until the bat alights on its roost.

Mating occurs in the autumn and it is thought that sperm are stored during hibernation with fertilisation occuring the following spring, but some authorities think a second mating in the spring results in fertilisation. The young are born between May and June. Each fertile female will only produce a single offspring each year. The young bat clings to the mother's undersurface where it feeds

49

Pipistrellus
Pipistrelle bat

PHYLUM CHORDATA

SUBPHYLUM VERTEBRATA [CRANIATA]

CLASS MAMMALIA

INFRACLASS EUTHERIA

ORDER CHIROPTERA

GENUS *Pipistrellus*

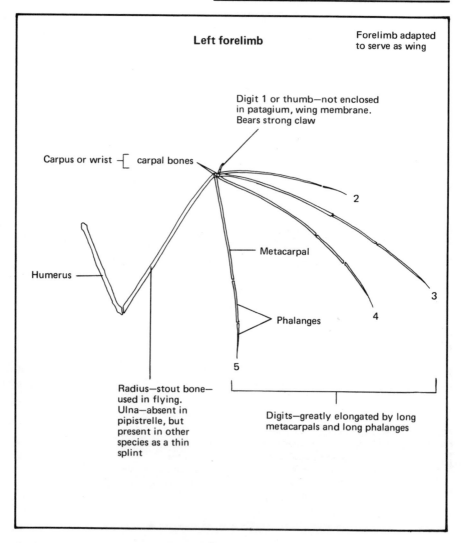

Left forelimb

Forelimb adapted to serve as wing

Digit 1 or thumb—not enclosed in patagium, wing membrane. Bears strong claw

Carpus or wrist —[carpal bones

2

Metacarpal

Humerus —

3

Phalanges   4

5

Radius—stout bone— used in flying. Ulna—absent in pipistrelle, but present in other species as a thin splint

Digits—greatly elongated by long metacarpals and long phalanges

from her. When the young bat becomes too large to be carried by the mother, it is left behind in the colony. Lactation lasts for 2—3 weeks at the end of which time the permanent teeth have erupted.

Bats are able to avoid obstacles while flying in the dark, by emitting ultra-sonic sounds which are reflected back to the ear. By this method of echo-location they are able to judge accurately the course in front of them.

Hibernation occurs before the end of October and lasts until the end of March, but it may be interrupted on mild nights.

PHYLUM CHORDATA	
SUBPHYLUM VERTEBRATA [CRANIATA]	
CLASS MAMMALIA	
INFRACLASS EUTHERIA	
ORDER PRIMATES	
GENUS *Pan*	

Pan
Chimpanzee

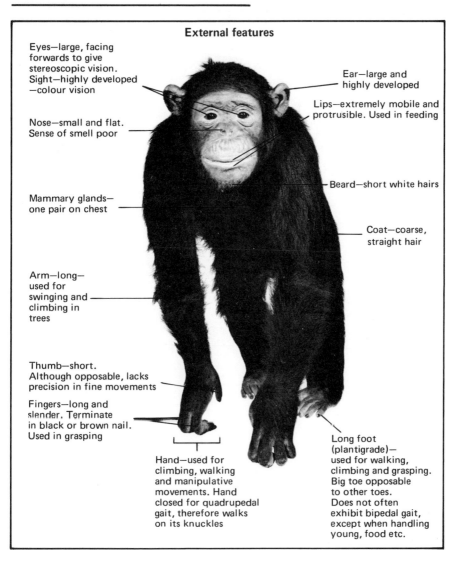

External features

Eyes—large, facing forwards to give stereoscopic vision. Sight—highly developed —colour vision

Nose—small and flat. Sense of smell poor

Mammary glands— one pair on chest

Arm—long— used for swinging and climbing in trees

Thumb—short. Although opposable, lacks precision in fine movements

Fingers—long and slender. Terminate in black or brown nail. Used in grasping

Hand—used for climbing, walking and manipulative movements. Hand closed for quadrupedal gait, therefore walks on its knuckles

Ear—large and highly developed

Lips—extremely mobile and protrusible. Used in feeding

Beard—short white hairs

Coat—coarse, straight hair

Long foot (plantigrade)— used for walking, climbing and grasping. Big toe opposable to other toes. Does not often exhibit bipedal gait, except when handling young, food etc.

Chimpanzees are found in West and Central Equatorial Africa. They live in groups of families or bands and travel over wide areas in accordance with weather conditions and availability of food. They are arboreal in habit, but spend much time on the ground. Sleep and rest periods are spent in trees where they construct nest-like sleeping places. They are intelligent animals and there is much communication amongst themselves by smell, tactile, visual and auditory signals. They are herbivorous, eating fruit, leaves and young shoots, but they are reputed to be

51

Pan
Chimpanzee

PHYLUM CHORDATA

SUBPHYLUM VERTEBRATA [CRANIATA]

CLASS MAMMALIA

INFRACLASS EUTHERIA

ORDER PRIMATES

GENUS *Pan*

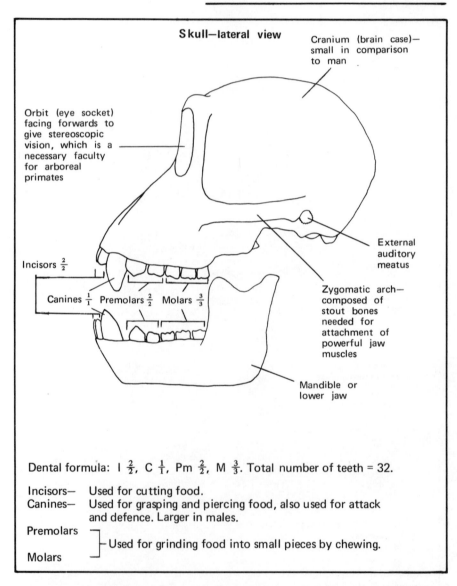

Skull—lateral view

Cranium (brain case)—small in comparison to man

Orbit (eye socket) facing forwards to give stereoscopic vision, which is a necessary faculty for arboreal primates

External auditory meatus

Incisors $\frac{2}{2}$

Canines $\frac{1}{1}$ Premolars $\frac{2}{2}$ Molars $\frac{3}{3}$

Zygomatic arch—composed of stout bones needed for attachment of powerful jaw muscles

Mandible or lower jaw

Dental formula: I $\frac{2}{2}$, C $\frac{1}{1}$, Pm $\frac{2}{2}$, M $\frac{3}{3}$. Total number of teeth = 32.

Incisors— Used for cutting food.

Canines— Used for grasping and piercing food, also used for attack and defence. Larger in males.

Premolars ⎫
⎬ Used for grinding food into small pieces by chewing.
Molars ⎭

carnivorous, since they have been seen to catch and eat small mammals and insects.

The females menstruate at monthly intervals after which they will receive males. Mating occurs throughout the year but there seem to be indications of a definite breeding season. Gestation lasts 7—9 months. The females give birth to one, sometimes two infants which have smooth brown skin and little hair. The young are

PHYLUM CHORDATA

SUBPHYLUM VERTEBRATA [CRANIATA]

CLASS MAMMALIA

INFRACLASS EUTHERIA

ORDER PRIMATES

GENUS *Pan*

Pan
Chimpanzee

Whole skeleton

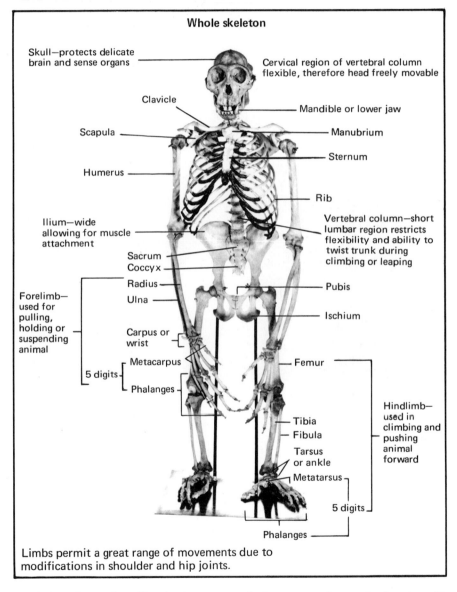

Skull—protects delicate brain and sense organs

Cervical region of vertebral column flexible, therefore head freely movable

Clavicle

Mandible or lower jaw

Scapula

Manubrium

Sternum

Humerus

Rib

Ilium—wide allowing for muscle attachment

Vertebral column—short lumbar region restricts flexibility and ability to twist trunk during climbing or leaping

Sacrum
Coccyx

Radius

Pubis

Ulna

Ischium

Forelimb—used for pulling, holding or suspending animal

Carpus or wrist

Femur

Metacarpus

5 digits

Phalanges

Hindlimb—used in climbing and pushing animal forward

Tibia

Fibula

Tarsus or ankle

Metatarsus

5 digits

Phalanges

Limbs permit a great range of movements due to modifications in shoulder and hip joints.

suckled by the mother. Weaning is a very gradual process and extends after the milk dentition has been completed. Parental care is excellent and other members of the band, males and females alike, take interest in and care for the young. Sexual maturity—attained 8 -12 years.

Oryctolagus
Rabbit

PHYLUM CHORDATA

SUBPHYLUM VERTEBRATA [CRANIATA]

CLASS MAMMALIA

INFRACLASS EUTHERIA

ORDER LAGOMORPHA

GENUS *Oryctolagus*

External features

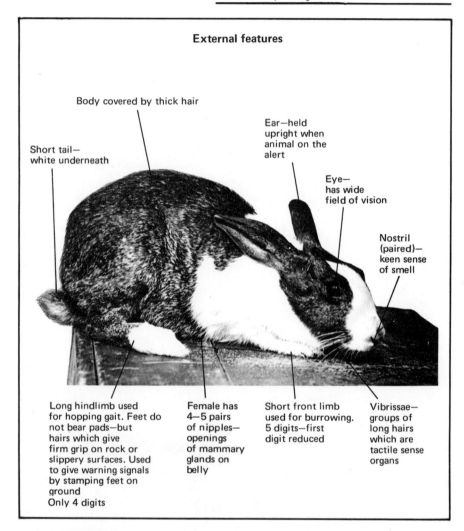

Body covered by thick hair

Ear—held upright when animal on the alert

Eye— has wide field of vision

Short tail— white underneath

Nostril (paired)— keen sense of smell

Long hindlimb used for hopping gait. Feet do not bear pads—but hairs which give firm grip on rock or slippery surfaces. Used to give warning signals by stamping feet on ground. Only 4 digits

Female has 4—5 pairs of nipples— openings of mammary glands on belly

Short front limb used for burrowing. 5 digits—first digit reduced

Vibrissae— groups of long hairs which are tactile sense organs

Rabbits are gregarious animals which are found in most lowland areas. They live in warrens, which are a series of interconnecting tunnels excavated by the rabbits in light, dry soil.

Rabbits are crepuscular or nocturnal in their habits and the adults do not move far from the warren in order to feed. The females, or does, may enlarge the warren by excavating new tunnels, which are incorporated into the original warren. Also the young rabbits may move into different warrens.

They feed mainly on grass, but also eat leaves of other plants and young trees, causing much damage to agricultural land and crops. They are known to eat moist,

PHYLUM CHORDATA	***Oryctolagus***
SUBPHYLUM VERTEBRATA [CRANIATA]	Rabbit
CLASS MAMMALIA	
INFRACLASS EUTHERIA	
ORDER LAGOMORPHA	
GENUS *Oryctolagus*	

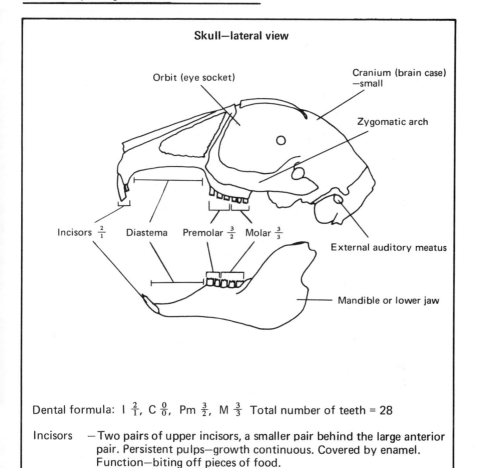

Skull—lateral view

Orbit (eye socket)

Cranium (brain case) —small

Zygomatic arch

Incisors $\frac{2}{1}$ Diastema   Premolar $\frac{3}{2}$ Molar $\frac{3}{3}$

External auditory meatus

Mandible or lower jaw

Dental formula: I $\frac{2}{1}$, C $\frac{0}{0}$, Pm $\frac{3}{2}$, M $\frac{3}{3}$ Total number of teeth = 28

Incisors —Two pairs of upper incisors, a smaller pair behind the large anterior pair. Persistent pulps—growth continuous. Covered by enamel. Function—biting off pieces of food.

Canines —Absent.

Premolars —Possess sharp transverse ridges—used for cutting food.

Molars —Upper teeth bite outside lower teeth.

early morning and late evening pellets which are high in bacterially produced B. vitamins.

Breeding occurs between January and June and mating takes place in the burrows, rarely in the field. Gestation lasts four weeks. The doe makes a nest to receive her litter, which ranges from three to seven individuals. The young are suckled by the mother for three weeks, at the end of which time they can feed themselves. The young rabbits will breed in their first year and their mother is usually pregnant very soon after giving birth to the previous litter which accounts for the very high reproductive rate.

Oryctolgus
Rabbit

PHYLUM CHORDATA

SUBPHYLUM VERTEBRATA [CRANIATA]

CLASS MAMMALIA

INFRACLASS EUTHERIA

ORDER LAGOMORPHA

GENUS *Oryctolagus*

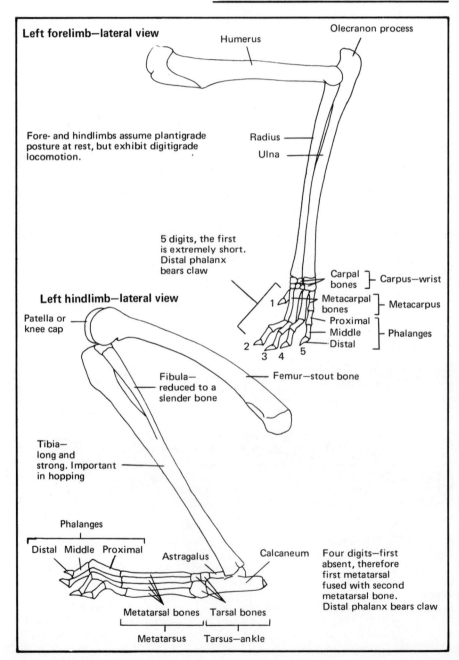

Left forelimb—lateral view

Humerus

Olecranon process

Fore- and hindlimbs assume plantigrade posture at rest, but exhibit digitigrade locomotion.

Radius

Ulna

5 digits, the first is extremely short. Distal phalanx bears claw

Carpal bones — Carpus—wrist

Metacarpal bones — Metacarpus

Proximal
Middle — Phalanges
Distal

1 2 3 4 5

Left hindlimb—lateral view

Patella or knee cap

Fibula— reduced to a slender bone

Femur—stout bone

Tibia— long and strong. Important in hopping

Phalanges

Distal Middle Proximal

Astragalus

Calcaneum

Four digits—first absent, therefore first metatarsal fused with second metatarsal bone. Distal phalanx bears claw

Metatarsal bones   Tarsal bones

Metatarsus   Tarsus—ankle

PHYLUM CHORDATA	***Rattus***

PHYLUM CHORDATA
SUBPHYLUM VERTEBRATA [CRANIATA]
CLASS MAMMALIA
INFRACLASS EUTHERIA
ORDER RODENTIA
GENUS *Rattus*

Rattus
Black rat

External features

Long cylindrical body— very flexible due to flexibility of spine

Ear—well developed pinna. Hearing is very acute

Eye— keen sight

Vibrissae— groups of long hairs which are tactile sense organs

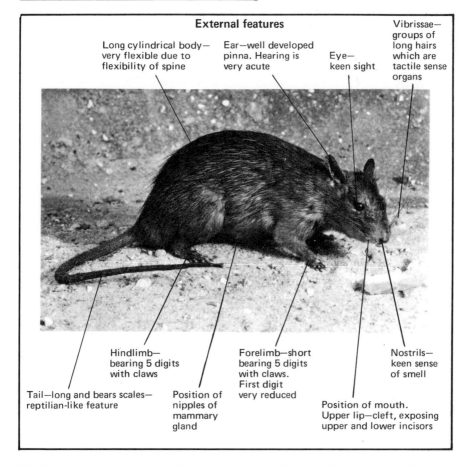

Hindlimb— bearing 5 digits with claws

Forelimb—short bearing 5 digits with claws. First digit very reduced

Nostrils— keen sense of smell

Tail—long and bears scales— reptilian-like feature

Position of nipples of mammary gland

Position of mouth. Upper lip—cleft, exposing upper and lower incisors

The black rat is not a native of Britain, but was 'imported' with cargoes from Southern Asia. It is now found all over the world in close proximity to human habitation. The rats live in burrows in large buildings, such as warehouses, in cellars, in sewers and ships, but are generally confined to ports.

They feed on almost anything—food debris, stored products and grain, to which they travel along definite pathways from their burrows.

Breeding seems to occur all the year round with females producing four to five litters a year, each one consisting of about seven young. Gestation lasts three weeks. The young when born are put in a nest. They are naked, blind and quite helpless. Lactation lasts about three weeks, by which time the young can cope with a mixed diet. The young rats attain sexual maturity between four and six months, which accounts for the very high reproductive rate.

Rats are of great economic importance because of their scavenging habits on man's resources, and because they transmit pathogenic organisms.

Rattus
Black rat

PHYLUM CHORDATA

SUBPHYLUM VERTEBRATA [CRANIATA]

CLASS MAMMALIA

INFRACLASS EUTHERIA

ORDER RODENTIA

GENUS *Rattus*

Skull—lateral view

Milk teeth—rudiments disappear before birth

Diastema—long gap between incisors and cheek teeth. Skin can be folded in across it during gnawing to prevent food from going into back of mouth

Orbit (eye socket)

Cranium (brain case)

Zygomatic arch— large

Enamel

Posterior part of tooth covered by thin layer of cement

Incisors $\frac{1}{1}$

Molars $\frac{3}{3}$

External auditory meatus—leads to middle ear

Mandible or lower jaw

Dental formula: I $\frac{1}{1}$, C $\frac{0}{0}$, Pm $\frac{0}{0}$, M $\frac{3}{3}$. Total number of teeth = 16

Incisors— Large with open pulp cavities so that they grow continuously. Ends of teeth are chisel shaped, due to differential wear between enamel and cement. Used for cutting and gnawing.

Canines— Completely absent.

Premolars— Absent.

Molars— Several cusps on each molar which are free from enamel— differential wear between enamel and dentine produces an uneven surface on teeth, which are used for grinding food. Upper teeth bite inside lower teeth.

PHYLUM CHORDATA

SUBPHYLUM VERTEBRATA [CRANIATA]

CLASS MAMMALIA

INFRACLASS EUTHERIA

ORDER RODENTIA

GENUS *Rattus*

Rattus
Black rat

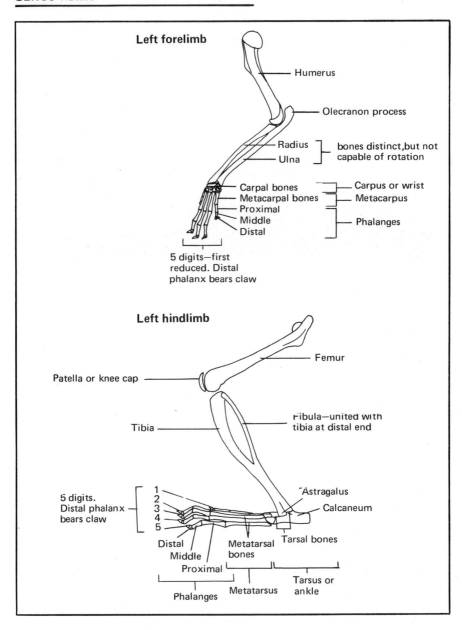

Left forelimb

- Humerus
- Olecranon process
- Radius
- Ulna
- bones distinct, but not capable of rotation
- Carpal bones
- Metacarpal bones
- Proximal
- Middle
- Distal
- Carpus or wrist
- Metacarpus
- Phalanges

5 digits—first reduced. Distal phalanx bears claw

Left hindlimb

- Femur
- Patella or knee cap
- Fibula—united with tibia at distal end
- Tibia
- 5 digits. Distal phalanx bears claw
- 1 2 3 4 5
- Astragalus
- Calcaneum
- Distal
- Middle
- Proximal
- Metatarsal bones
- Tarsal bones
- Phalanges
- Metatarsus
- Tarsus or ankle

Forelimbs and hindlimbs assume a plantigrade posture at rest, but exhibit digitigrade locomotion.

Globicephala
Pilot whale, Blackfish or Caa'ing whale

PHYLUM CHORDATA

SUBPHYLUM VERTEBRATA [CRANIATA]

CLASS MAMMALIA

INFRACLASS EUTHERIA

ORDER CETACEA

GENUS *Globicephala*

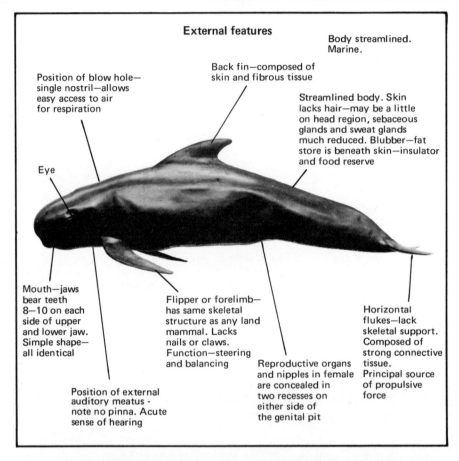

External features

Body streamlined. Marine.

Position of blow hole—single nostril—allows easy access to air for respiration

Back fin—composed of skin and fibrous tissue

Streamlined body. Skin lacks hair—may be a little on head region, sebaceous glands and sweat glands much reduced. Blubber—fat store is beneath skin—insulator and food reserve

Eye

Mouth—jaws bear teeth 8—10 on each side of upper and lower jaw. Simple shape—all identical

Flipper or forelimb—has same skeletal structure as any land mammal. Lacks nails or claws. Function—steering and balancing

Horizontal flukes—lack skeletal support. Composed of strong connective tissue. Principal source of propulsive force

Reproductive organs and nipples in female are concealed in two recesses on either side of the genital pit

Position of external auditory meatus - note no pinna. Acute sense of hearing

The Pilot Whale is found around the northern coasts of Britain. They swim around in large schools which are composed of up to a hundred individuals.

They feed on fish and cephalopods in the form of squid and cuttlefish. The whale's body bears scars as witness to these encounters.

Breeding occurs throughout the year. Mating takes place in the sea during the autumn and gestation lasts 13—16 months. The birth of the young whale or calf also occurs in the water. It either swims to the surface in order to breathe or is pushed up by its mother. The calf is suckled by its mother for about a year. During the lactatory period the nipples are obvious and protrude from the ventral surface. While feeding the calf frequently surfaces for air.

Whales are both excellent swimmers and divers and are able to remain submerged in deep water for several minutes. They are also able to make use of echolocation for detecting objects and appear to communicate with each other by audible signals.

PHYLUM CHORDATA	*Globicephala*
SUBPHYLUM VERTEBRATA [CRANIATA]	Pilot whale
CLASS MAMMALIA	
INFRACLASS EUTHERIA	
ORDER CETACEA	
GENUS *Globicephala*	

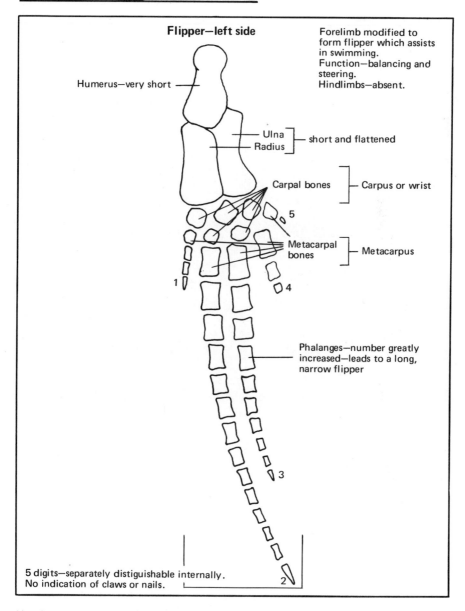

Flipper—left side

Forelimb modified to form flipper which assists in swimming. Function—balancing and steering. Hindlimbs—absent.

Humerus—very short

Ulna
Radius — short and flattened

Carpal bones — Carpus or wrist

5

Metacarpal bones — Metacarpus

1

4

Phalanges—number greatly increased—leads to a long, narrow flipper

3

5 digits—separately distiguishable internally.
No indication of claws or nails.

2

Hand narrowed by loss of one digit (rudimentary). Movement of the flipper is effected at the shoulder joint—no movement between rest of the bones of the flipper.

61

Felis
Cat

PHYLUM CHORDATA

SUBPHYLUM VERTEBRATA [CRANIATA]

CLASS MAMMALIA

INFRACLASS EUTHERIA

ORDER CARNIVORA

GENUS *Felis*

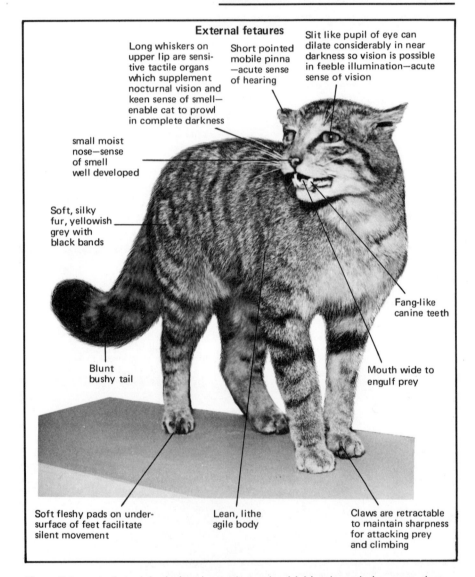

External fetaures

Long whiskers on upper lip are sensitive tactile organs which supplement nocturnal vision and keen sense of smell—enable cat to prowl in complete darkness

Short pointed mobile pinna —acute sense of hearing

Slit like pupil of eye can dilate considerably in near darkness so vision is possible in feeble illumination—acute sense of vision

small moist nose—sense of smell well developed

Soft, silky fur, yellowish grey with black bands

Fang-like canine teeth

Blunt bushy tail

Mouth wide to engulf prey

Soft fleshy pads on undersurface of feet facilitate silent movement

Lean, lithe agile body

Claws are retractable to maintain sharpness for attacking prey and climbing

The wild cat is found in isolated woods, rocky highlands and the open deer moors of Scotland, Wales and the highlands of Europe. The den or nest is usually built in an isolated rock crevice, and is lined by grass or heather.

The wild cat avoids man completely. It hunts hares, rabbits, grouse, birds, poultry and lambs, mainly at night. The prey is devoured where it is caught. The keen senses and silent, swift locomotion are modifications for a predacious life.

PHYLUM CHORDATA
SUBPHYLUM VERTEBRATA [CRANIATA]
CLASS MAMMALIA
INFRACLASS EUTHERIA
ORDER CARNIVORA
GENUS *Felis*

Felis
Cat

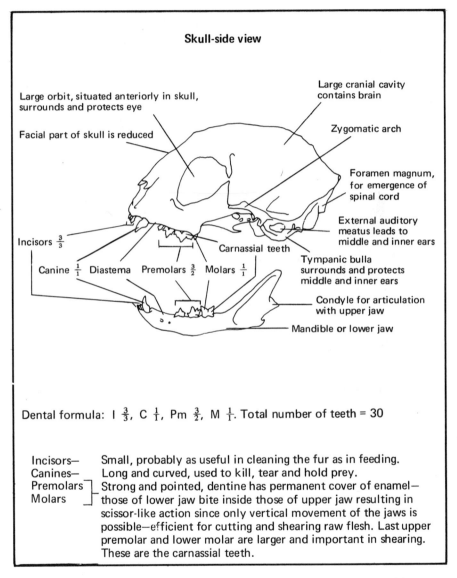

Skull-side view

Large cranial cavity contains brain

Large orbit, situated anteriorly in skull, surrounds and protects eye

Zygomatic arch

Facial part of skull is reduced

Foramen magnum, for emergence of spinal cord

External auditory meatus leads to middle and inner ears

Incisors $\frac{3}{3}$

Carnassial teeth

Tympanic bulla surrounds and protects middle and inner ears

Canine $\frac{1}{1}$ Diastema Premolars $\frac{3}{2}$ Molars $\frac{1}{1}$

Condyle for articulation with upper jaw

Mandible or lower jaw

Dental formula: I $\frac{3}{3}$, C $\frac{1}{1}$, Pm $\frac{3}{2}$, M $\frac{1}{1}$. Total number of teeth = 30

Incisors—	Small, probably as useful in cleaning the fur as in feeding.
Canines—	Long and curved, used to kill, tear and hold prey.
Premolars	Strong and pointed, dentine has permanent cover of enamel— those of lower jaw bite inside those of upper jaw resulting in scissor-like action since only vertical movement of the jaws is possible—efficient for cutting and shearing raw flesh. Last upper premolar and lower molar are larger and important in shearing. These are the carnassial teeth.
Molars	

The wild cat is a solitary animal, males and females only come together at mating. After a period of gestation lasting 68 days, as many as five kittens are born in the nest and are guarded ferociously by the mother. The kittens themselves spit and fight. After a period of suckling, the kittens leave the nest and live for up to sixteen years.

63

Felis
Cat

PHYLUM CHORDATA

SUBPHYLUM VERTEBRATA [CRANIATA]

CLASS MAMMALIA

INFRACLASS EUTHERIA

ORDER CARNIVORA

GENUS *Felis*

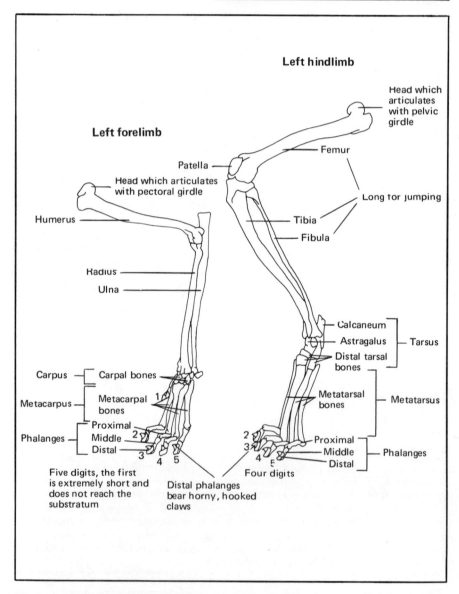

The cat exhibits a digitigrade mode of progression—palms and soles are raised off the ground—so only the digits are in contact with the substratum. This enables the cat to move swiftly and stealthily. The soft pads underneath the feet and retractable claws also facilitate silent movement.

PHYLUM CHORDATA	***Halichoerus***

PHYLUM CHORDATA

SUBPHYLUM VERTEBRATA [CRANIATA]

CLASS MAMMALIA

INFRACLASS EUTHERIA

ORDER PINNIPEDIA

GENUS *Halichoerus*

Halichoerus
Grey seal

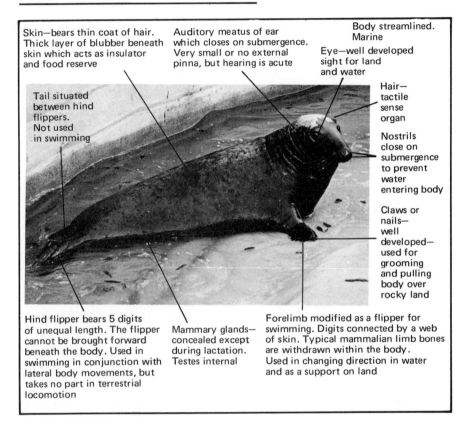

Skin—bears thin coat of hair. Thick layer of blubber beneath skin which acts as insulator and food reserve

Auditory meatus of ear which closes on submergence. Very small or no external pinna, but hearing is acute

Body streamlined. Marine

Eye—well developed sight for land and water

Tail situated between hind flippers. Not used in swimming

Hair—tactile sense organ

Nostrils close on submergence to prevent water entering body

Claws or nails—well developed—used for grooming and pulling body over rocky land

Hind flipper bears 5 digits of unequal length. The flipper cannot be brought forward beneath the body. Used in swimming in conjunction with lateral body movements, but takes no part in terrestrial locomotion

Mammary glands—concealed except during lactation. Testes internal

Forelimb modified as a flipper for swimming. Digits connected by a web of skin. Typical mammalian limb bones are withdrawn within the body. Used in changing direction in water and as a support on land

The grey seal is a marine animal, but it comes onto rocky shores round the coast to breed, moult or fish in shore waters. When breeding they form distinct breeding colonies away from human habitation, so they tend to be found on isolated islands.

They are carnivores and feed on molluscs, crustaceans and fish, which are often swallowed whole.

Mating occurs immediately after the females have finished lactation which lasts only 2—4 weeks. It is essential it occurs soon after birth as it is the only time females are aggregated in large numbers. Mating occurs either on land or in the water. [There is a delay of about five months before implantation occurs]. After mating, the breeding colony breaks up towards the middle of November, and the seals return to the sea to feed and recuperate. The following year in September, the breeding colony reassembles. The bulls arrive first and show certain territorial behaviour. The young pups are born with a white, silky coat which they lose in a series of moults before they attain the typical adult coat. A pregnant seal will produce one pup annually. The pup is suckled by its mother for 2—4 weeks after which time it must fend for itself and out of necessity take to the water to find food.

65

Halichoerus
Grey seal

PHYLUM CHORDATA

SUBPHYLUM VERTEBRATA [CRANIATA]

CLASS MAMMALIA

INFRACLASS EUTHERIA

ORDER PINNIPEDIA

GENUS *Halichoerus*

Skull—lateral view

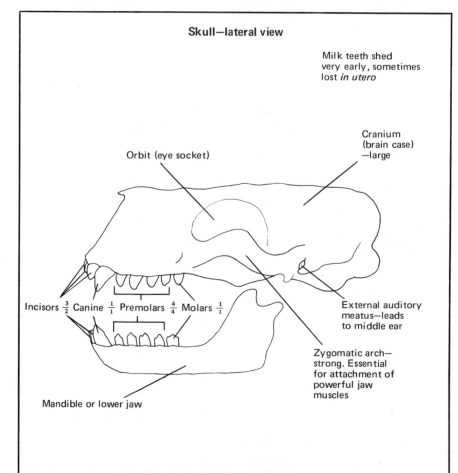

Milk teeth shed very early, sometimes lost *in utero*

Cranium (brain case) —large

Orbit (eye socket)

Incisors $\frac{3}{2}$ Canine $\frac{1}{1}$ Premolars $\frac{4}{4}$ Molars $\frac{1}{1}$

External auditory meatus—leads to middle ear

Zygomatic arch— strong. Essential for attachment of powerful jaw muscles

Mandible or lower jaw

Dental formula: I $\frac{3}{2}$, C $\frac{1}{1}$, Pm $\frac{4}{4}$, M $\frac{1}{1}$. Total number of teeth = 34

Incisors— Used for grasping and biting prey.

Canines— Used for piercing and grasping prey.

Premolars ⌐ Often known as post-canines, since they are almost indistinguish-
Molars ⌐ able from one another. Used for tearing prey. Those found in the upper jaw have single cusps, but those in the lower jaw are almost tricuspid. Prey is usually swallowed whole, but if it is too large it is torn into smaller, more manageable pieces by these teeth.

PHYLUM CHORDATA

SUBPHYLUM VERTEBRATA [CRANIATA]

CLASS MAMMALIA

INFRACLASS EUTHERIA

ORDER PINNIPEDIA

GENUS *Halichoerus*

Halichoerus
Grey seal

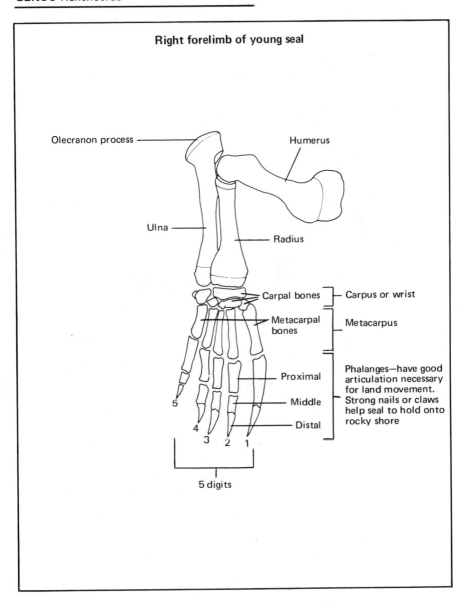

Right forelimb of young seal

Olecranon process

Humerus

Ulna

Radius

Carpal bones — Carpus or wrist

Metacarpal bones — Metacarpus

Proximal

Middle

Distal

Phalanges—have good articulation necessary for land movement. Strong nails or claws help seal to hold onto rocky shore

5

4

3 2 1

5 digits

Humerus, radius and ulna are short and stout. The hand is broad and flat. The bones provide attachment and a point of insertion for the powerful limb muscles which are used for changing direction in the water and as an aid in the terrestrial 'looping' locomotion.

Equus
Horse

PHYLUM CHORDATA

SUBPHYLUM VERTEBRATA [CRANIATA]

CLASS MAMMALIA

INFRACLASS EUTHERIA

ORDER PERISSODACTYLA

GENUS *Equus*

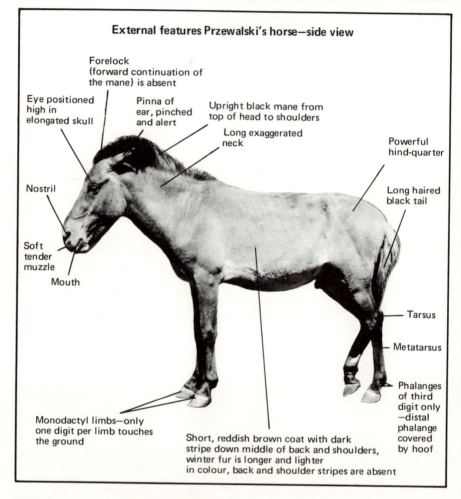

External features Przewalski's horse—side view

Forelock (forward continuation of the mane) is absent

Eye positioned high in elongated skull

Pinna of ear, pinched and alert

Upright black mane from top of head to shoulders

Long exaggerated neck

Powerful hind-quarter

Nostril

Long haired black tail

Soft tender muzzle

Mouth

Tarsus

Metatarsus

Phalanges of third digit only —distal phalange covered by hoof

Monodactyl limbs—only one digit per limb touches the ground

Short, reddish brown coat with dark stripe down middle of back and shoulders, winter fur is longer and lighter in colour, back and shoulder stripes are absent

The Mongolian wild horse (Przewalski's horse) survives in small numbers on the grassy plains of central Asia.

These wild horses live in herds led by an old stallion. They graze by day and migrate to new pastures. During the mating season the stallions fight amongst themselves using their teeth and hooves. The gestation period is long and lasts from 330—365 days, but mares do not breed every year. The foal has a very well developed coat and can stand and run immediately after birth, but is still suckled on milk from the inguinal mammary glands. Przewalski's horse lives up to 28 years in captivity.

Domestic species of horse are of value as race horses, cart horses and as meat in certain countries.

PHYLUM CHORDATA	***Equus***
SUBPHYLUM VERTEBRATA [CRANIATA]	Horse
CLASS MAMMALIA	
INFRACLASS EUTHERIA	
ORDER PERISSODACTYLA	
GENUS *Equus*	

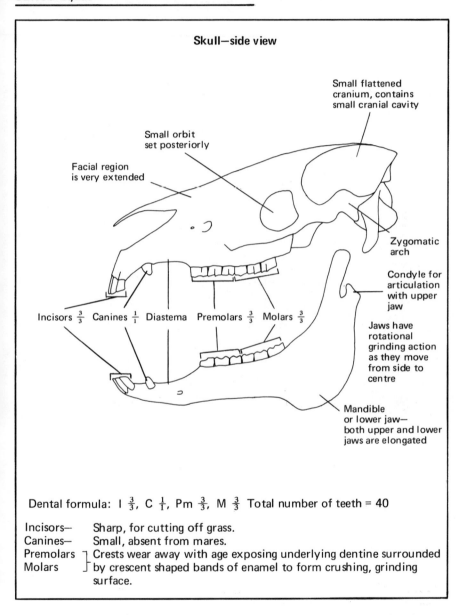

Skull—side view

Small flattened cranium, contains small cranial cavity

Small orbit set posteriorly

Facial region is very extended

Zygomatic arch

Condyle for articulation with upper jaw

Incisors $\frac{3}{3}$ Canines $\frac{1}{1}$ Diastema Premolars $\frac{3}{3}$ Molars $\frac{3}{3}$

Jaws have rotational grinding action as they move from side to centre

Mandible or lower jaw— both upper and lower jaws are elongated

Dental formula: I $\frac{3}{3}$, C $\frac{1}{1}$, Pm $\frac{3}{3}$, M $\frac{3}{3}$ Total number of teeth = 40

Incisors—	Sharp, for cutting off grass.
Canines—	Small, absent from mares.
Premolars ⎤ Molars ⎦	Crests wear away with age exposing underlying dentine surrounded by crescent shaped bands of enamel to form crushing, grinding surface.

The teeth and jaw action of the horse are adapted for a herbivorous diet. So is the extended caecum of the gut containing micro-organisms which break down cellulose to fatty acids.

Equus
Horse

PHYLUM CHORDATA

SUBPHYLUM VERTEBRATA [CRANIATA]

CLASS MAMMALIA

INFRACLASS EUTHERIA

ORDER PERISSODACTYLA

GENUS *Equus*

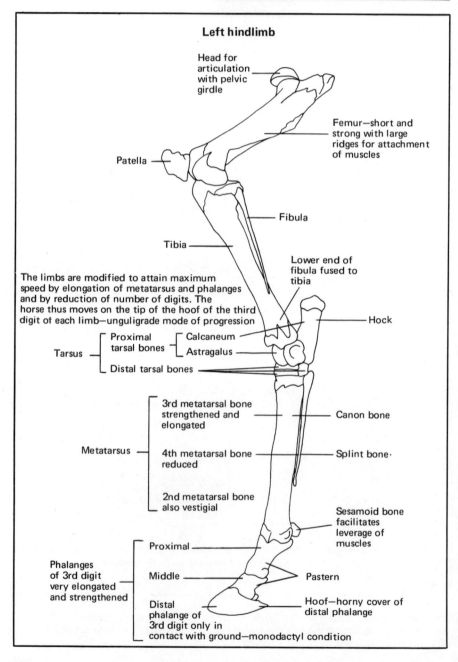

Left hindlimb

Head for articulation with pelvic girdle

Femur—short and strong with large ridges for attachment of muscles

Patella

Fibula

Tibia

Lower end of fibula fused to tibia

Hock

The limbs are modified to attain maximum speed by elongation of metatarsus and phalanges and by reduction of number of digits. The horse thus moves on the tip of the hoof of the third digit of each limb—unguligrade mode of progression

Tarsus — Proximal tarsal bones — Calcaneum / Astragalus

Distal tarsal bones

Metatarsus —
3rd metatarsal bone strengthened and elongated — Canon bone

4th metatarsal bone reduced — Splint bone

2nd metatarsal bone also vestigial

Sesamoid bone facilitates leverage of muscles

Phalanges of 3rd digit very elongated and strengthened
Proximal

Middle — Pastern

Distal phalange of 3rd digit only in contact with ground—monodactyl condition

Hoof—horny cover of distal phalange

70

PHYLUM CHORDATA	***Ovis***
SUBPHYLUM VERTEBRATA [CRANIATA]	Sheep
CLASS MAMMALIA	
INFRACLASS EUTHERIA	
ORDER ARTIODACTYLA	
GENUS *Ovis*	

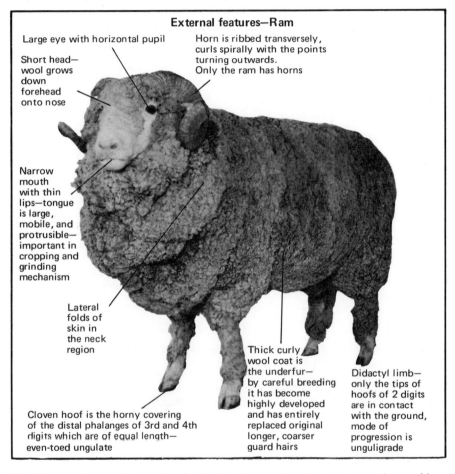

External features—Ram

Large eye with horizontal pupil

Short head—wool grows down forehead onto nose

Horn is ribbed transversely, curls spirally with the points turning outwards. Only the ram has horns

Narrow mouth with thin lips—tongue is large, mobile, and protrusible—important in cropping and grinding mechanism

Lateral folds of skin in the neck region

Cloven hoof is the horny covering of the distal phalanges of 3rd and 4th digits which are of equal length—even-toed ungulate

Thick curly wool coat is the underfur—by careful breeding it has become highly developed and has entirely replaced original longer, coarser guard hairs

Didactyl limb—only the tips of hoofs of 2 digits are in contact with the ground, mode of progression is unguligrade

The Merino sheep originated in Spain, but is now found throughout the world, particularly in Australia, where it is the most important fine wool producing breed. It is a hardy animal, suited to dry climates, sure-footed and able to run on rocky slopes out of the range of predators.

Sheep live in herds and are active by day, cropping grass and other plants. The ruminant, four chambered stomach is modified for the digestion of plant material by micro-organisms as there is no enzyme which digests cellulose.

Breeding is seasonal with mating occuring in the autumn. After a period of 150 days gestation, 1 or 2 lambs are born. They are well developed with fur and hooves and are able to run as soon as they are born. They return frequently to the ewe for suckling.

Sheep live for about fourteen years and are of economic value for both wool and meat.

Ovis
Sheep

PHYLUM CHORDATA

SUBPHYLUM VERTEBRATA [CRANIATA]

CLASS MAMMALIA

INFRACLASS EUTHERIA

ORDER ARTIODACTYLA

GENUS *Ovis*

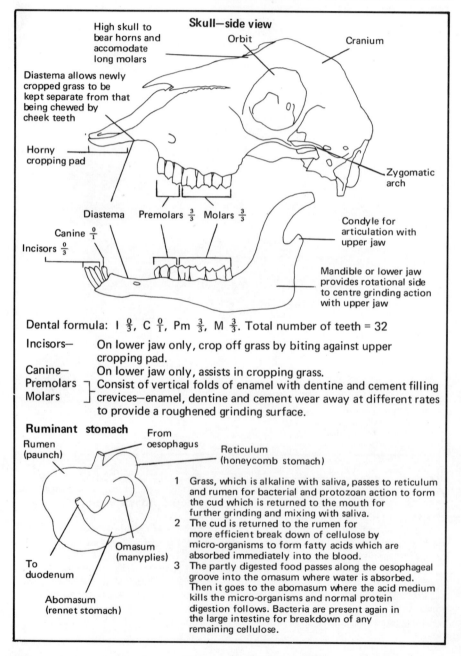

Skull—side view

High skull to bear horns and accomodate long molars

Orbit

Cranium

Diastema allows newly cropped grass to be kept separate from that being chewed by cheek teeth

Horny cropping pad

Zygomatic arch

Diastema   Premolars $\frac{3}{3}$ Molars $\frac{3}{3}$

Canine $\frac{0}{1}$

Incisors $\frac{0}{3}$

Condyle for articulation with upper jaw

Mandible or lower jaw provides rotational side to centre grinding action with upper jaw

Dental formula: I $\frac{0}{3}$, C $\frac{0}{1}$, Pm $\frac{3}{3}$, M $\frac{3}{3}$. Total number of teeth = 32

Incisors— On lower jaw only, crop off grass by biting against upper cropping pad.

Canine— On lower jaw only, assists in cropping grass.

Premolars ⌉ Consist of vertical folds of enamel with dentine and cement filling
Molars ⌋ crevices—enamel, dentine and cement wear away at different rates to provide a roughened grinding surface.

Ruminant stomach

Rumen (paunch)

From oesophagus

Reticulum (honeycomb stomach)

Omasum (manyplies)

To duodenum

Abomasum (rennet stomach)

1 Grass, which is alkaline with saliva, passes to reticulum and rumen for bacterial and protozoan action to form the cud which is returned to the mouth for further grinding and mixing with saliva.

2 The cud is returned to the rumen for more efficient break down of cellulose by micro-organisms to form fatty acids which are absorbed immediately into the blood.

3 The partly digested food passes along the oesophageal groove into the omasum where water is absorbed. Then it goes to the abomasum where the acid medium kills the micro-organisms and normal protein digestion follows. Bacteria are present again in the large intestine for breakdown of any remaining cellulose.

PHYLUM CHORDATA

SUBPHYLUM VERTEBRATA [CRANIATA]

CLASS MAMMALIA

INFRACLASS EUTHERIA

ORDER ARTIODACTYLA

GENUS *Ovis*

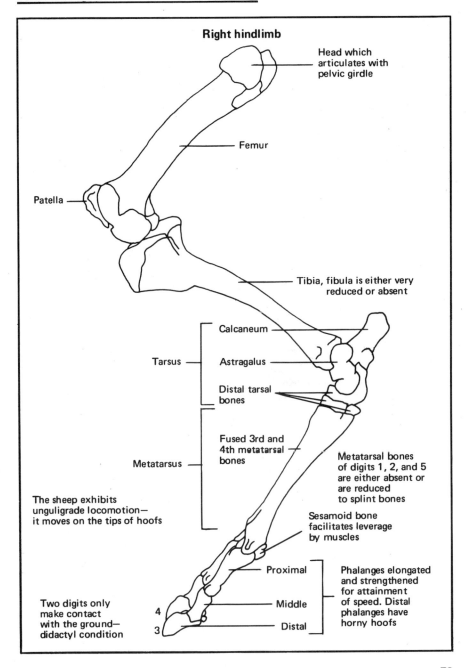

Right hindlimb

Head which articulates with pelvic girdle

Femur

Patella

Tibia, fibula is either very reduced or absent

Calcaneum

Tarsus

Astragalus

Distal tarsal bones

Fused 3rd and 4th metatarsal bones

Metatarsus

Metatarsal bones of digits 1, 2, and 5 are either absent or are reduced to splint bones

The sheep exhibits unguligrade locomotion— it moves on the tips of hoofs

Sesamoid bone facilitates leverage by muscles

Proximal

Phalanges elongated and strengthened for attainment of speed. Distal phalanges have horny hoofs

Two digits only make contact with the ground— didactyl condition

4

Middle

3

Distal

Experimental approach to the vertebrates

PREPARATION OF SLIDES

a) Temporary preparations
1. Fixation
2. Staining
3. Mounting

b) Permanent preparations
1. Fixation
2. Staining
3. Differentiation
4. Dehydration
5. Clearing
6. Mounting

(See explanation of stages below)

FIXATION	The preparation is fixed or killed, thereby keeping it in its life-like state. The most frequently used fixative is 70% alcohol.
STAINING	This enables different parts of the preparation to be distinguished and differentiated from each other, because they take up the stain to varying extents. HAEMATOXYLIN is used for staining vertebrate tissues, e.g. muscle, cartilage, tendon, etc. BORAX CARMINE may also be used for vertebrate tissues, but tends to be taken up in excess and is then difficult to remove.
DIFFERENTIATION	Acid alcohol will remove excess stain, so that the detailed structures are distinguishable under the micr-scope.
DEHYDRATION	Water must be removed from the preparation. Alcohol is a dehydrating agent. The preparation is taken up through the alcohols from 70% → 90% absolute alcohol. This stage must be gradual as immersion in a strong solution of alcohol could harm a delicate preparation.
CLEARING	The preparation must be cleared to remove traces of alcohol and to ensure that the tissue is transparent. Xylol or clove oil are clearing agents. If dehydration has not been taken to completion, the clearing agent will turn cloudy or milky.
MOUNTING	Permanent preparations are mounted in a resinous material such as Canada balsam. Temporary preparations are mounted in water or glycerine.

STAINING SCHEDULES

a) Permanent Preparations

BORAX CARMINE
1. 50% alcohol for two minutes.
2. Stain in borax carmine for five to ten minutes.
3. Wash in 50% alcohol.
4. Wash in 70% alcohol.
5. Differentiate in acid alcohol and check under the microscope.
6. Wash in 70% alcohol.
7. Rinse in 90% alcohol for one minute.
8. Transfer to absolute alcohol for two rinses of five minutes each. Keep covered.
9. Clear in xylol for at least ten minutes.
10. Mount in Canada balsam (do not stir up the Canada balsam in its bottle, as this will introduce air bubbles. The problem can be overcome by using Canada balsam from a tube.)

HAEMATOXYLIN
1. 70% alcohol for two minutes (if tissue is alive, leave for five to ten minutes.)
2. Stain in haematoxylin for five to fifteen minutes.
3. Rinse in 70% alcohol.
4. Differentiate in acid alcohol and check under the microscope.
5. Wash in 70% alcohol.
6. Run under tap water until blue (haematoxylin turns red in acid and blue in alkaline conditions.)
7. Wash in 70% alcohol.
8. Counterstain in eosin for ten seconds (this is optional).
9. Rinse in 90% alcohol for one minute.
10. Transfer to absolute alcohol for two rinses of five minutes each. Keep covered.
11. Clear in xylol for at least ten minutes.
12. Mount in Canada balsam.

b) Temporary Preparations

PREPARATION OF A BLOOD SMEAR
Use aseptic technique (sterilise skin with alcohol and use a sterile lancet or mounted needle.)
1. Place a drop of fresh blood on a clean slide and make a thin smear by allowing drop of blood to spread along the edge of a second slide and pushing second slide along the first, inclined at an angle of 45°. (See below).

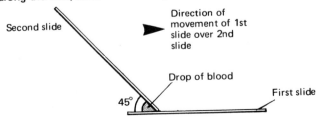

Second slide

Direction of movement of 1st slide over 2nd slide

Drop of blood

First slide

45°

2. Dry the smear of blood by waving the slide about in the air.
3. Add Leishman's stain and leave for about thirty seconds.
4. Add an equal quantity of distilled water.
5. Rock the slide to mix the water and stain together and leave for ten minutes.
6. Drain off the excess liquid and rinse with distilled water.
7. Blot with filter paper, air-dry and examine under the microscope.

N.B. To make a permanent preparation, mount in Canada balsam or green euparal, when the blood smear has been completely air dried.

PREPARATION OF A SMEAR OF CHEEK EPITHELIAL CELLS
1. Scrape squamous epithelial cells from the inside of the mouth, using a wooden spatula and make a thin smear on a slide.
2. Add methylene blue stain immediately. (Do not let the cells dry up at any time).
3. Leave for a few minutes.
4. Place a coverslip on the smear and examine under the microscope.

FOOD TESTS

1. CARBOHYDRATES
Found in bread, potatoes, cereals and sweet foods.
Composed of carbon, hydrogen and oxygen.
The hydrogen and oxygen occur in a 2 : 1 ratio, as in a molecule of water.
The general formula for a carbohydrate is $C_x(H_2O)_y$.

There are three types of carbohydrate.
a) *Monosaccharides* Simple sugars with one sugar molecule.
   e.g. glucose, fructose and galactose.
   The general formula for glucose is $C_6H_{12}O_6$.

b) *Disaccharides* Double sugars with two sugar molecules.
   e.g. Glucose + Glucose → Maltose (Malt Sugar).
       Glucose + Fructose → Sucrose (Cane Sugar).
       Glucose + Galactose → Lactose (Milk Sugar).
The general formula for sucrose is $C_{12}H_{22}O_{11}$.
N.B. Monosaccharides and disaccharides dissolve in water and are sweet to the taste.

c) *Polysaccharides* Multiple sugars made up of many monosaccharide units.
   linked together to form a very large molecule.
   e.g. starch, glycogen and cellulose.
The general formula for the polysaccharides is $[C_6H_{10}O_5]_n$.
Polysaccharides are insoluble in water and are used as reserves of carbohydrate because they can be stored by the organism. Plants store their carbohydrate as starch, whilst animals utilise glycogen.

Tests for Sugars

A. FEHLING'S TEST
To 1 cm^3 of glucose solution add sufficient Fehling's solution to give the contents of the test tube a blue colour. (Fehling's solution is made by mixing equal volumes of Fehling's 1 and 2 together).
Boil for a few minutes.

An orange-red precipitate of copper oxide indicates the presence of a reducing sugar. The copper sulphate in the Fehling's solution has, therefore, been reduced to copper oxide. Hence glucose is often referred to as a reducing sugar.

Sucrose is a non-reducing sugar, so it must be broken down into its constituent monosaccharide units before the Fehling's test can be carried out.
If sucrose is boiled with dilute hydrochloric acid, it will be hydrolysed to glucose and fructose.
The addition of solid sodium bicarbonate will neutralise the acid, and when no more effervescence (bubbling due to carbon dioxide) occurs, the bicarbonate has completely neutralised the acid.
Now test for a reducing sugar as before.

B. BENEDICT'S TEST
To 2 cm^3 of Benedict's solution, add a few drops of glucose solution, in a test tube.
Place the test tube in a beaker containing boiling water for 2 to 5 minutes.
A greenish-yellow precipitate indicates the presence of glucose.
It may be necessary to cool rapidly the solution in the test tube, to bring down the precipitate. This can be done by placing the test tube under running cold water from the tap.

To test for sucrose, dilute hydrochloric acid and solid sodium bicarbonate must be added as before, to convert the disaccharide into the monosaccharides. Then proceed with the Benedict's test as for glucose.

Test for Starch

IODINE TEST
To 1 cm^3 of a solution of fresh starch add a few drops of iodine solution. This may also be done using a white tile on which a drop of starch and iodine solution are mixed together. A blue-black colour indicates the presence of starch.

2. FATS
Found as almost pure fat in butter, lard and margarine.
Also found in foods such as milk, nuts and fatty meat.
Composed of carbon, hydrogen and oxygen, combined together in different proportions from the carbohydrates.

Fats are made up of fatty acid and glycerol.
Three molecules of fatty acid and one molecule of glycerol $\rightarrow$ one molecule of fat.

77

Stearic acid is a fatty acid found in beef fat.

Similarly oleic acid is found in olive oil and
Palmitic acid is found in coconut oil.

Tests for Fats

A. TRANSLUCENT STAIN TEST
Fats are greasy in nature and if smeared onto paper or cloth they will leave a permanent translucent stain which does not dry out.

B. EMULSION TEST
Take 2 cm^3 of ethyl alcohol in a test tube and add a few drops of olive oil. Shake well so that the oil mixes in well with the alcohol.
Now add 2 cm^3 of cold water.
The presence of fat is indicated by the formation of small droplets or globules of fat in suspension.
These droplets give a milky appearance.

3. PROTEINS
Found in meat, fish, milk, cheese and eggs.
Composed of carbon, hydrogen, oxygen, nitrogen and sometimes sulphur and phosphorous.
They are molecules which have a high molecular weight.
They are made up of amino acids linked together to form long chains.
They are made up of amino acids linked together by peptide linkages (CO.NH) to form long chains.
The general formula for an amino acid is:

$$H_2N-\underset{\underset{R}{|}}{\overset{\overset{H}{|}}{C}}-COOH$$

The amino acid has a basic amino group $-NH_2$, at one end, and an acidic carboxyl group $-COOH$, at the other end. Because of its acidic and basic properties an amino acid is said to be amphoteric.
R varies depending on the particular amino acid, e.g. glycine is the simplest amino acid where R is an atom of hydrogen.

Tests for Proteins

A. BIURET TEST
To 2 cm^3 of a solution of egg albumen add 3 cm^3 of sodium hydroxide and shake the test tube well.
Then add one drop of copper sulphate.
A violet colour indicates the presence of protein.

B. MILLON'S TEST
To 5 cm^3 of egg albumen solution add 2 cm^3 of Millon's reagent.
A white precipitate may form.
Boil for one minute.
A brick-red precipitate indicates the presence of protein. .

D. XANTHOPROTEIC TEST
To 3 cm^3 of egg albumen solution add 1 cm^3 of concentrated nitric acid.
A white precipitate should appear.
Boil until the solution is clear yellow.
Cool and add concentrated ammonium hydroxide to give an orange colour.
This indicates the presence of protein.

D. NINHYDRIN TEST
To 1 cm^3 of egg albumen solution add 2 drops of Ninhydrin.
Heat until the solution turns blue. This indicates the presence of protein.
N.B.　The Ninhydrin test indicates the presence of amino (NH_2) groups in proteins.
　　　The Biuret test indicates the presence of peptide linkages between
　　　amino acids.
The Xanthoproteic test indicates the presence of specific amino acids.
e.g. tyrosine, tryptophane and phenylalanine.
The Millon's test indicates the presence of the amino acid tyrosine.

EXPERIMENT TO ILLUSTRATE THE PRINCIPLES OF PAPER CHROMATOGRAPHY

Apparatus

2 amino acids: L—glutamic acid
　　　　　　　　L—proline
Ethanol
Butan—1—ol/Acetic Acid/Water solvent
Ninhydrin
Acetone
Pyridine
Whatman No. 1 or 2 chromatography paper or
　blotting paper or filter paper
100 cm^3 measuring cylinder
Two 10 cm^3 measuring cylinders
1 litre tall beaker or a coffee jar and lid
Shallow dish
Hot air blower or hair drier
2 glass rods, scissors, sellotape, pencil and ruler

Method
1.　Take a 10 cm^3 measuring cylinder and dissolve glutamic acid in 1 cm^3 ethanol to produce a saturated solution. Repeat with proline, using another measuring cylinder.
　　N.B. Both amino acids are soluble in ethanol.

tip of a glass rod, and stir it into a drop of iodine solution on a white tile.

N.B.   It may be easier to place several rows of iodine drops on the white tile before starting the experiment.

6.   Note the time taken for the blue-black colouration of iodine and starch to disappear completely. This time is the chromic period, and the end point, when iodine remains yellow in colour, is the achromic point.

Questions

1.   How long was the chromic period?
2.   What has happened to the starch at the achromic point?
3.   Why is it advantageous to carry out the experiment at a temperature of 37° C?
4.   What control experiment could be used and why is a control necessary?

INVESTIGATION OF THE ACTION OF PTYALIN OVER A RANGE OF TEMPERATURES

Apparatus   (as in the previous experiment)

Method

1.   Take a test tube and add 10 cm^3 of starch solution and leave in a water bath at 40° C until the temperature of the starch solution is also 40° C.
2.   Add 1 cm^3 of diluted saliva and start the stop watch immediately.
     N.B. To maintain a constant temperature, place the Bunsen burner at the edge of the wire gauze rather than directly underneath the water bath.
3.   At intervals of 30 seconds withdraw a drop of solution and test for the presence of starch with iodine, as in the previous experiment.
     N.B.   If the chromic period is less than 2 minutes, dilute the enzyme with more distilled water. If it is longer than 20 minutes, a more concentrated enzyme preparation must be made, or the experiment will take too long.
4.   Repeat this procedure using new solutions of starch and saliva, at temperatures of 30°, 50°, 60° and 100° C.
5.   Note the achromic point for each of these tests and plot the results as a graph of time in minutes against temperature in degrees Centigrade. Time is plotted on the ordinate or vertical axis and temperature on the abscissa or horizontal axis.
     N.B.   The non-variable data is placed usually along the abscissa, whilst the data that has been acquired experimentally is placed along the ordinate.
     N.B.   The time taken to reach the achromic point is a measurement of the rate of reaction. When plotting the results, use the reciprocal of the length of chromic period, i.e. $\frac{1}{Time}$.

Questions

1.   What is the name for the shape of the graph?
2.   What is the optimum temperature at which ptyalin functions best?
3.   What happens to the enzyme at 100° C. and why are enzymes thermolabile?

INVESTIGATION OF THE ACTION OF PTYALIN OVER A RANGE OF pH

Apparatus As in the previous experiment.
Also a range of solutions of pH 5, 6, 7, 8 and 9.

Method
1. Take a test tube and add 8 cm^3 of starch and 2 cm^3 of a buffer solution of pH 5. Then add 1 cm^3 of diluted saliva.
2. Place the test tube in a water bath at 37° C. and at intervals of 30 seconds remove a drop of solution and test for the presence of starch with iodine indicator.
   Note the length of the chromic period at this pH value.
3. Repeat with pH values of 6, 7, 8 and 9, carrying out each of these tests at a temperature of 37° C.

Questions
1. Is there an optimum pH for ptyalin?
2. Does the enzyme prefer an acid, neutral or alkaline medium?
3. Considering that the site of ptyalin's activity is in the mouth, should one expect this particular enzyme to be absolutely specific about the pH of the medium it works in?

INVESTIGATION OF THE EFFECTS OF VARIOUS SALTS ON THE ACTION OF PTYALIN

Apparatus As in the previous experiment
Also a 0·5% solution of the following salts:
Sodium chloride
Potassium chloride
Sodium nitrate
Potassium nitrate

Method
1. Take a test tube and add 9 cm^3 of starch solution and 1 cm^3 of 0·5% sodium chloride. Then add 1 cm^3 of diluted saliva.
2. Place the test tube in a water bath at 37° C. and at intervals of 30 seconds remove a drop of solution and test for the presence of starch with iodine.
   Note the length of the chromic period.
3. Repeat using 1 cm^3 of 0·5% potassium chloride, sodium nitrate and potassium nitrate.
   Carry out all these tests at a temperature of 37° C.

Questions
1. In which salt solution did digestion of starch by ptyalin occur most rapidly?
2. Conversely, in which salt solution did enzyme activity proceed most slowly?
3. From the results is it possible to deduce which salts or ions activate the enzyme and which retard it?
4. Some ions may have no direct effect on the enzyme. Which are they?
5. What control should be carried out in this experiment?

EXPERIMENT TO ILLUSTRATE OSMOSIS

Apparatus   Concentrated sucrose solution
                                 Distilled water
                                 Visking tubing (approximately 2 cms. wide)
                                 Elastic band
                                 Thistle funnel or capillary tubing
                                 Large and small beaker

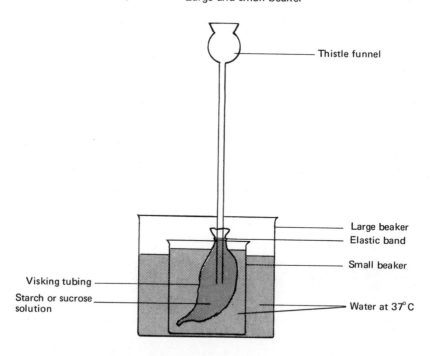

Thistle funnel

Large beaker
Elastic band

Small beaker

Visking tubing

Starch or sucrose
solution

Water at 37° C

Method
1. Take a 15 cm. strip of visking tubing and moisten under the tap to soften it.
2. Tie a knot in one end and then fill with a concentrated sucrose solution.
3. Attach the other end of the visking tubing to a thistle funnel using an elastic band wrapped round many times to provide a complete seal (see diagram).
   N.B.   If the seal between the glass tube and the visking tube is not good, the liquid will escape at this junction.
4. Place the visking tubing in a small beaker full of distilled water and put the small beaker in a large beaker containing water at 37° C.
5. Adjust the thistle funnel or capillary tubing so that it is held in an upright position. It may be necessary to support it.
6. Mark the level of the column of liquid in the glass tubing, leave for a period of time and then check the level and note any change in its position.

Questions
1. How can the results be explained in terms of osmosis?
2. Visking tubing is functioning as a semi-permeable membrane. What is a semi-permeable membrane?
3. What analogies can be made between this simple apparatus and a cell?
4. Could this experiment be used to measure the osmotic pressure of the sucrose solution?
5. Why is it necessary to carry out the experiment at a temperature of 37° C.

INVESTIGATION INTO THE PERMEABILITY OF A MEMBRANE

Apparatus

10% starch solution
10% glucose solution
Iodine/potassium iodide solution
Benedict's reagent
Visking tubing (approximately 2 cms. in width)
A white tile, pipettes and paper clips
Bunsen burner, tripod, wire gauze and asbestos mat
A small beaker or boiling tube and a large beaker
Thermometer

Method
1. Take a 15 cm. strip of visking tubing and moisten under the tap to soften it.
2. Tie a knot in one end and fill with equal quantities of starch and glucose solutions using a pipette.
3. Seal the other end with a paper clip or bulldog clip.
4. Wash the outside of the visking tubing to remove any traces of starch or glucose that may have spilt over.
5. Fill a small beaker or a boiling tube with just sufficient warm tap water to cover the visking tubing and place the tubing in the water.
6. Place the boiling tube or beaker in a large beaker containing water at 37° C.
   N.B.   It is necessary to carry out the experiment at this temperature in order that it will work in a fairly short period of time.
7. Now test some of the water surrounding the visking tubing for the presence of starch or glucose, using the iodine and Benedict's tests respectively.
   A sample of water may be withdrawn with a clean pipette.
8. Repeat these tests again after 25 minutes and record the results.

Questions
1. Knowing the visking tubing acts as a membrane which is permeable to some substances, but not to others, how can the results obtained in the experiment be explained?
2. What analogies can be made with the process of digestion in the gut?
3. Why is digestion an essential process in the animal?

INVESTIGATION INTO THE TOLERANCE OF RED BLOOD CELLS TO CHANGES IN EXTERNAL OSMOTIC PRESSURE

Apparatus

Fresh blood
0·85% sodium chloride solution
2% sodium chloride solution
Distilled water
3 test tubes and a test tube rack
3 glass rods and 3 pipettes
3 microscope slides and cover slips
Microscope

Method

1. Take 3 clean test tubes and place 1 cm^3 of 0·85% sodium chloride in the first test tube, 1 cm^3 of 2% sodium chloride in the second and 1 cm^3 of distilled water in the third. Arrange these test tubes in the rack in this order.
   N.B.   Use a clean pipette for each solution.
2. Obtain a sample of blood from the thumb using the aseptic technique—sterilise the skin with cotton wool soaked in alcohol and then pierce the skin at base of nail with a lancet. It may help to constrict the tip of the thumb first to ensure a good flow of blood from the wound.
   Defibrinated blood may also be used. This has the advantage of not clotting when exposed to air.
3. Place a drop of blood in each of the 3 test tubes, using a clean glass rod each time, and stir well.
   N.B.   This blood need not be diluted beforehand.
   Do not allow the blood to stand for long because this will cause clotting.
4. Take a few drops of the liquid containing blood and 0·85% sodium chloride from the first test tube and place on a clean microscope slide.
   Cardfully lay a cover slip on top and view under the microscope, using the high power objective. Note the appearance of the red blood cells and draw them.
   N.B.   0·85% sodium chloride is of the same osmotic pressure as the contents within the red blood cells.
5. Repeat, using the solutions in the other 2 test tubes, and again record the results in the form of annotated drawings.

Questions

1. Knowing that 0·85% sodium chloride is of the same osmotic pressure as the contents of the red blood cell itself, would you expect to see any changes in the red blood cells?
2. In which solution did haemolysis or rupture of the red blood cells occur and why should the erythrocytes burst?
3. In which osmotic medium did the red blood cells shrink and shrivel up?
   It should be possible to see corrugation of the cell membrane, giving a spiky appearance to the red blood cell. Explain the responses on the part of the red blood cells to this osmotic medium.
4. Would you expect the tolerance of the red blood cells to external fluctuations in osmotic pressure to be very great?

BLOOD TYPES

There are 4 main blood groups: A, B, AB and O.

This classification of blood types is based upon the fact that blood possesses antigen A and antigen B factors.

The red blood cells or erythrocytes carry the antigens and these substances are capable of reacting with a corresponding antibody present in the blood plasma. When this occurs the red blood cells agglutinate or clump together

Normally, the antibody corresponding to a particular antigen does not occur in the blood of the same individual, but in another individual. However, if the blood from this other individual, of a different blood type, is introduced into the circulatory system of the first, agglutination will occur.

Blood group A possesses antigen A and this will react with the corresponding antibody A, but not with antibody B. Similarly, blood group B has antigen B. Blood group AB has both antigen A and B; whereas blood group O has neither antigen present.

It is possible to determine the blood group of an individual by bringing the blood into contact with a blood serum known to contain one of the two antigens. It can then be clearly seen if the blood agglutinates or not.

The table below indicates the possible reactions:

Blood group	Reaction of blood of unknown type to:	
	Antibody A ≡ Blood Group B.	Antibody B ≡ Blood Group A.
A	x	√
B	√	x
AB	x	x
O	√	√

√ = compatability of blood and serum.
   Therefore no agglutination occured.

x = non-compatability of blood and serum.
   Therefore agglutination occured.

N.B.   Blood group AB is a universal recipient; whilst blood group O is a universal donor.

The latter is not true strictly speaking, because there are so many other antigens present in the blood, that in practice no single blood type could be completely compatible with all other variations in blood type.

DETERMINATION OF BLOOD GROUPS

Apparatus

Blood of unknown blood group
Blood serum of anti-A and anti-B type
0·85% sodium chloride solution
Test tube
Microscope slide
Felt-tipped pen
2 pipettes and a glass rod

Method
1. Place 1 drop of "donor" blood in 1 cm^3 of 0·85% saline in a test tube and mix well using a glass rod.
2. Take a clean microscope slide and divide into two, labelling one side anti-A or blood group B, and the other side anti-B or blood group A.
3. Place a drop of serum of anti-A or anti-B on the appropriate half of the slide, using a clean pipette for each serum, to avoid contamination.
   N.B.   When blood clots, the red blood cells are held together leaving a clear straw-coloured fluid containing no red blood corpuscles. This is the serum.
4. On each drop of serum place a drop of the blood/saline mixture and leave for about 10 minutes, ensuring that it does not dry up.
5. Note the presence or absence of agglutinated red blood cells, and deduce the blood group.

EXPERIMENT TO SHOW CONTRACTION OF A MUSCLE FIBRE

Apparatus

Two muscle fibres from fresh lean meat
ATP solution (adenosine triphosphate)
1% glucose solution
Ringer's solution
Distilled water
Ice
Ruler
Syringe and pipette
Microscope slides

Method
1. Tease out individual muscle fibres from a piece of fresh lean meat. The fibres should be about 2 mm. in thickness.
2. Place one muscle fibre on a clean microscope slide and measure its length with a millimetre ruler.
   N.B.   Do not handle the muscle fibre any more than is absolutely necessary and ensure that it does not dry up. It should be kept moist with Ringer's solution.
3. Run 0·5 cm^3 of distilled water from a syringe onto the muscle fibre. Leave for one to two minutes and measure its length again.

4. Drain off the excess fluid and repeat the procedure using 0·5 cm^3 glucose solution.
5. Again drain off the excess fluid and add 0·5 cm^3 ATP solution. Note any change in the length of the muscle fibre.
   N.B.   To make up a solution of ATP, dissolve ·05 grams ATP in 5 cm^3 distilled water. This will be sufficient to fill ten syringes with 0·5 cm^3 ATP solution.
6. Take a new muscle fibre, measure its length, and then run 0·5 cm^3 ATP solution over it.
   Leave for one to two minutes and measure its length again.

Questions
1. Under what conditions does a muscle fibre contract?
2. Was there any difference in the extent of contraction of a muscle fibre when
   (a) Glucose plus ATP, ⎤
   (b) ATP solution   ⎦ was added?
3. Calculate the extent to which a muscle fibre shortens, as a percentage of its original length.

$$\frac{\text{Original length} - \text{Final length on shortening}}{\text{Original length}} \times 100 = \frac{\text{\% Contraction of a muscle fibre}}{}$$

INVESTIGATION OF CONTRACTION IN THE GASTROCNEMIUS MUSCLE OF THE FROG

   N.B.   This experiment involves pithing a frog and is therefore more suitable for class demonstration purposes. Chloroform cannot be used as this would interfere with the activities of the nerves and muscles.

   Apparatus   See diagram, page 90.

   Method
A. To obtain a Sciatic Gastrocnemius preparation
1. To dissect out the sciatic nerve and the gastrocnemius muscle, first remove the skin from the lower half of the pithed frog.
2. Cut through the Achilles tendon which runs round the heel of the foot and isolate the gastrocnemius muscle as far as the knee.
3. Cut through the tibia and fibula just below the knee.
4. Now isolate the gastrocnemius muscle from the other leg muscles, above the knee, and expose the sciatic nerve which runs between them.
5. To trace the sciatic nerve to its point of origin in the spinal cord cut a "window" in the lower back region of the frog.
6. When the 7th, 8th and 9th spinal nerves can be seen leaving the vertebral column on each side, cut transversely through the backbone and remove a small section of it, with the nerves attached.
7. Trace these three nerves into the leg where they fuse to form the sciatic plexus, and follow the sciatic nerve to within one centimetre of the knee.
   N.B.   Dissect out the nerve on one side only.

8. Cut through the femur near to the knee joint so that the entire nerve-muscle preparation can be removed from the frog.

    N.B. During the dissection the preparation must be lubricated with Ringer's solution to keep it moist.

    Handling should be kept to a minimum.

B. To set up the apparatus

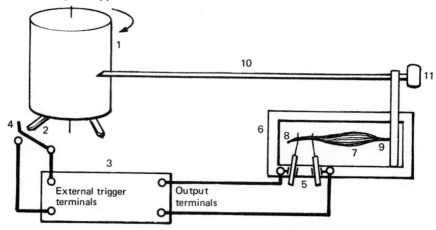

Key

1. Kymograph or drum	7. Gastrocnemius muscle
2. Spindle or contact arm	8. Sciatic nerve (7th, 8th and 9th spinal nerves)
3. Electrical stimulator	9. Achilles tendon
4. Contact maker	10. Recording lever and pointer
5. Electrode	11. After loading screw
6. Muscle bath containing Ringer's solution	

1. Arrange the kymograph and wrap the kymograph paper around the drum, securing it with glue or sellotape.

    N.B. This paper has a layer of soot deposited on it by revolving the drum in a flame of coal gas which has passed through benzene. Special graph paper may also be used and this has the advantage of being easier to handle. The trace of muscle contraction is recorded with ink on this paper.

2. Connect up the apparatus as shown in the diagram, so that the two electrodes link up with the output terminals of the stimulator, and the contact maker on the kymograph links up with the external trigger terminals of the stimulator.

3. Insert a pin through the knee joint and tie a piece of thread round the tendon, with a loop at the end.

4. Fix the muscle firmly to the floor of the muscle bath, by means of a pin

and attach the tendon to the arm of the recording lever by the loop at the end of the thread.

5. Arrange the two electrodes so that the sciatic nerve drapes over and is supported by them.

    N.B.   Do not stretch the nerve at anytime. When the experiment is not in progress, lower the electrodes so that the nerve lies in the Ringer's solution in the muscle bath. If it dries up it will cause spontaneous muscle twitches which will spoil the results recorded on the kymograph. However, during the experiment the nerve must be lifted well out of the Ringer's solution so that the electrical current is not short-circuited.

C.   To record an Isotonic twitch

1. Arrange the arm of the recording lever so that it is horizontal in position and so that the weight is taken by the muscle. Any adjustment that needs to be made can be done by tightening or loosening the thread which attaches the tendon to the base of the lever.

2. Tighten the "after loading" screw at the pivot point of the lever, so that the weight is just taken off the muscle. The muscle is now said to be after-loaded; previously it was free-loaded.

3. Set the spindle arms of the kymograph 180° apart and turn the drum so that one of the contact arms touches the contact maker at the base of the drum. At this point of contact the electrical circuit is completed and the muscle can be stimulated by electrical current, via the two electrodes.

4. Operate the strength control dial on the stimulator to produce an electrical stimulus sufficient to elicit a maximum contraction of the muscle.

    N.B.   Keep the recording lever away from the kymograph during this trial run.

5. Adjust the position of the drum on its spindle or axis, so that one of the two spindle arms points just after the join in the kymograph paper. This will ensure that the recordings of muscle contraction do not run over the join.

6. Place the pointer of the recording lever on the surface of the drum. It should press lightly and be horizontal in position.

    N.B.   The writing point must be trailing against the surface of the revolving drum and not leading into it. If it is pressing too lightly it will not record properly and must be moved closer to the drum.

7. Let the drum revolve once, so that the pointer can record a baseline. At this stage the muscle must not be stimulated and, therefore, the external trigger terminals should be disconnected to break the electrical circuit.

8. Now record two muscle twitches for just one revolution of the drum, at a medium speed, and then stop the drum.

    N.B.   Because the two spindle arms are 180° apart, the two recordings of muscle contraction will occur on opposite sides of the drum, equidistant from each other.

9. To record the actual point of stimulation, revolve the drum slowly by hand in the same direction as before and manually operate the contact maker with the spindle arm. At the point of contact the muscle will contract and as the drum is moving slowly, a single vertical line will be recorded.

    This marks the moment of stimulation of the sciatic nerve.

    Repeat for the trace on the other side of the drum.

10. Remove the recording lever and place a 100 or 50 cycles per second time-marker against the rotating drum, so that a time trace is recorded just below the baseline. The drum should be rotated at the same speed that was used for the recording of the muscle contractions.

    N.B.   The time trace is very important because it provides the units for the horizontal axis of the graph.

           The degree of movement of the recording lever provides the units for the vertical axis.

Questions

1. What is the difference between an isotonic and an isometric muscle contraction?
2. From the recorded muscle contraction, measure the following:
   a. Latent period
   b. Contraction period
   c. Relaxation period
   d. Height of contraction (distance muscle shortens)
   e. Velocity of shortening $= \dfrac{\text{Height of contraction}}{\text{Time}}$

    N.B.   The letters a, b, c, and d in this question correspond to the letters on the diagram.

Recorded Trace of Muscle Contraction

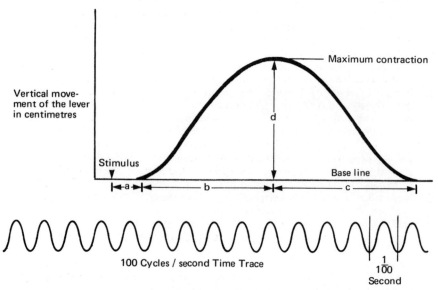

N.B.   This experiment can be used to illustrate summation and tetanus in a muscle by increasing the frequency and strength of the electrical stimulus applied to the muscle.

INVESTIGATION INTO THE RATE OF FROG HEART BEAT OVER A RANGE OF TEMPERATURES

Apparatus

Frog
Ringer's solution
Large beaker
Pipette
Bunsen burner, tripod, gauze, asbestos mat
Thermometer
Stopwatch

Method
1. Pith a frog
2. Open up the thoracic cavity by cutting through the sternum and exposing the heart.
3. Lubricate the heart with Ringer's solution at room temperature (20° C) using a pipette.
4. Ensure that the heart is beating normally and count the number of contractions of the ventricle per minute with a stopwatch.
5. Repeat using a range of temperatures from 35°, 45°, 55°, and 65° C.
6. Cool the heart down gradually using iced Ringer's solution and then count the number of heart beats in 10°, 5°, and 0° C Ringer's solution.
   N.B. It is not advisable to subject the heart to the lowest temperatures until the end of the experiment because the heart may stop altogether.
7. Plot a graph of the number of heart beats per minute against the temperature of the Ringer's solution.

Questions
1. How does the heart of a frog respond to an increase and then a decrease in temperature?
2. How can the results be related to the fact that the frog is a poikilothermic—cold-blooded—animal?
3. Was there any observable change in the rate at which the phase of contraction spread over the heart from auricles to ventricle, when the heart was subject to a lowering of temperature?

Glossary

AERIAL	—	able to fly.
AGGLUTINATE	—	clumping together of small particles to form larger masses which are usually precipitated.
ALVEOLI	—	tiny air sacs at the ends of bronchioles in mammalian lungs.
AMPHOTERIC	—	possessing both acidic and basic properties
ANUS	—	posterior opening of alimentary canal.
AQUATIC	—	living in water.
ARBOREAL	—	living in trees.
ARCHENTERON	—	cavity forming primitive gut in the embryonic gastrula.
ARTICULATION	—	a joint or movement at a joint.
ASTRAGALUS	—	the upper bone of the foot on which the tibia rests.
ATRIUM	—	area enclosed by down growths of the body wall in Uro-chordates and Cephalochordates.
AUTOTOMY	—	self fracture.
BILATERAL SYMMETRY	—	symmetrical about a median plane.
BIPEDAL GAIT	—	movement on two feet.
BRILLE	—	transparent covering over a snake's eye.
CALCANEUM	—	heel bone.
CARNIVOROUS	—	feeding on the flesh of other animals.
CAUDAL	—	related to tail.
CEREBELLUM	—	part of brain concerned with co-ordination of movement.
CHROMATOPHORE	—	pigment cell.
CLEIDOIC	—	shelled egg in which embryo develops bathed in an aquatic medium surrounded by embryonic membranes.
CLOACA	—	common cavity into which open the alimentary, urinary and genital canals in birds, reptiles, most fish and monotremes.
COELOMATE	—	triploblastic animal with space or coelom present in the mesoderm separating it into outer somatic mesoderm applied to the body wall, and inner splanchnic mesoderm applied to the gut wall. The coelom connects with the exterior via gono-ducts and excretory ducts.
CREPUSCULAR	—	active in the twilight.
DENTAL FORMULA	—	is an accepted means of expressing the arrangement of teeth in a mammalian skull. The types of teeth are given symbols: I = incisors, C = canines, Pm = premolars, M = molars. The upper and lower lines refer to the upper and lower jaw of one half of the skull.
DIASTEMA	—	gap between two teeth.
DIGITIGRADE	—	digits in contact with ground.
DIPHYODONT	—	two sets of teeth, a milk set and a permanent set.
ECHOLOCATION	—	means of detecting obstacles by reflection of high frequency sounds produced by animal.
EFFERVESCENCE	—	giving off gas from a liquid—visible as bubbling.
EGG TOOTH	—	embryonic tooth of some reptiles and birds—for breaking shell.
EMBRYONIC MEMBRANES	—	extra embryonic tissue which surrounds and protects the embryo in reptiles, birds and mammals.
ENDOSKELETON	—	internal skeletal tissue.

EPIGLOTTIS	— fold of cartilage at back of tongue which covers glottis when swallowing.
EXOSKELETON	— skeletal tissue covering the animal.
EXTERNAL AUDITORY MEATUS	— tunnel in skull leading to the middle ear.
FERTILISATION	— fusion of nuclear material of two cells or gametes during sexual reproduction.
FOSSORIAL	— living in burrows, adapted for burrowing.
FRUGIVOROUS	— feeding on fruit.
GANGLION	— mass of nerve cell bodies.
GESTATION	— length of pregnancy.
GILL CLEFTS	— slits perforating the pharyngeal wall.
HAEMOLYSIS	— rupture of blood cells, releasing contents.
HERBIVOROUS	— feeding on plant material.
HETEROCERCAL	— lobes of tail asymmetrical.
HETERODONT	— teeth of different types.
HIBERNATION	— to spend winter in a dormant state.
HOMEOTHERMIC	— ability to maintain a constant body temperature.
HOMOCERCAL	— lobes of tail symmetrical.
HOMODONT	— teeth of the same type.
HYDROLYSIS	— interaction with water.
HYDROSTATIC ORGAN	— gas-filled organ derived from gut which alters animal's depth in water.
IMPLANTATION	— the attachment of a fertilised ovum to uterine wall.
INSECTIVOROUS	— feeding on insects.
INTROMITTENT ORGAN	— male copulatory organ.
LACTATION	— period during which the mammary glands secrete milk for nourishment of young.
LARVA	— free living embryo often concerned with dispersal which changes to adult by metamorphosis.
MARSUPIAL	— mammal with a pouch for development of foetus after birth.
MESONEPHRIC KIDNEY	— middle kidney, functional in adult fish and amphibia.
METAMERICALLY SEG-MENTED	— serial division of body into segments, each having same basic structure and origin.
METAMORPHOSIS	— rapid change in form.
METANEPHRIC KIDNEY	— hind kidney, functional in adult reptiles, birds and mammals.
MONOPHYODONT	— having only one set of teeth.
NIPPLE	— teat of mammary gland.
NOTOCHORD	— internal, dorsal axial skeletal rod in chordates, which may be replaced wholly or partly by a vertebral column.
OMNIVOROUS	— feeding on both animal and vegetable material.
OPERCULUM	— gill cover in bony fish.
OVIPAROUS	— eggs laid which hatch outside the body of the mother.
OVOVIVIPAROUS	— eggs hatch out in the body of the mother and the young are born alive or hatch out very soon after egg laying.
PATAGIUM	— wing membrane of bat.

Glossary

PECTORAL	—	related to chest or breast.
PENIAL	—	related to the penis
PELAGIC	—	living in surface waters.
PENTADACTYL	—	five digit limb, forms basic pattern of limbs in most amphibians, reptiles, birds and mammals.
PELVIC	—	related to the pelvis.
PERITONEUM	—	serous membrane lining body cavity and surrounding viscera.
PITCHING	—	downward movement of head into the sea.
PLACENTAL	—	furnished with placenta, an organ which attaches embryo to maternal uterus and serves to nourish it.
PLACOID	—	platelike.
PLANKTON	—	floating or drifting organisms found in the sea.
PLANTIGRADE	—	whole of lower surface of foot or hand in contact with ground.
POIKILOTHERMIC	—	cold blooded. Body temperature varies with external temperature.
PROTONEPHRIDIA	—	simple tubular excretory organs derived from ectoderm. Found in Cephalochordates.
PULP	—	soft core of tooth containing blood vessels and nerves.
QUADRUPEDAL GAIT	—	movement on four feet.
ROLLING	—	rotating about a vertical plane.
RUDIMENTARY	—	exhibiting only the basic elements.
SEBACEOUS GLANDS	—	sebum secreting glands.
SEBUM	—	substance with waterproofing and lubricating properties.
SESSILE	—	attached directly to substratum.
SPIRACLE	—	modified gill cleft.
SPIRAL VALVE	—	structure which serves to increase surface area in intestine of cartilaginous fish.
STEREOSCOPIC	—	three dimensional.
SUBCAUDAL	—	beneath the tail.
SUCKLE	—	to feed on milk produced by maternal mammary glands.
TERRESTRIAL	—	living on land.
TETRAPOD	—	four footed animal.
THERMOLABILE	—	tending to decompose on being heated.
TRAGUS	—	projection on outer ear or pinna.
TRIPLOBLASTIC	—	consisting of three layers: ectoderm, mesoderm and endoderm.
TUNICIN	—	material similar to cellulose found in the test of Urochordates.
TYMPANUM	—	ear drum.
UNGULIGRADE	—	only hoofs in contact with ground.
URETER	—	duct conveying excretory products from metanephric kidney.
VENT	—	common opening of alimentary, urinary and genital canals, or opening of anus.
VIBRISSAE	—	long, stiff hairs acting as tactile sense organs.
VIVIPAROUS	—	young being developed in close association with mother up to the time of birth.
WEANING	—	change of diet from maternally produced milk to other foods.
YAWING	—	deviation from a straight course.